ALASKA

NORTH TO THE FUTURE — VOLUME II

ALASKA

NORTH TO THE FUTURE – VOLUME II

The Publisher thanks the Governor's Office and the Alaska Department of Commerce and Economic Development for their support of this significant statewide project.

The Publisher also wishes to thank the Alaska State Legislature for its continuing support of this important series of economic development books.

Additionally, the Publisher would like to thank the National Oceanic and Atmospheric Administration, the Anchorage School District, and the Alaska State Chamber of Commerce.

The Publisher thanks Gail Phillips, Jim Clark, and other individuals too numerous to mention, for their unwavering support for this important project series.

Lastly, the Publisher thanks those individuals who contributed their inspiring thoughts at the beginning of each chapter of Part One, in order of their appearance: Peter Leathard, Jeff Staser, Ethel Lund, Governor William J. Sheffield, Mary Pignalberi, Drue Pearce, Norman Vaughan, and Fred Reeder.

Copyright 2003 by Wyndham Publications, Incorporated.
Printed in Canada.

ALL RIGHTS RESERVED. No part of this book may be used or reproduced in any manner whatsoever without written permission, except in the case of brief quotations embodied in critical articles or reviews. For information, mail your request to: Wyndham Publications, Incorporated; P.O. Box 45; Kirkland, Washington; 98083-0045. All information contained in this publication is accurate to the best knowledge of the Publisher.

Library of Congress Catalog Card Number: 2003112640
Library of Congress Information:
ALASKA: NORTH TO THE FUTURE, VOLUME II
Author: Ms. Heidi Bohi
Contributing Writer: Ms. Nancy Halverson
Major Contributing Photographers: Wyndham Images, AeroMap U.S.
Senior Vice President: Mr. Kim A. Halverson
Vice President of Production and Editor: Ms. Nancy Leichner
Dust Jacket Illustration: Ms. Megan Hatch, Service High School, Winner 2002 Wyndham Publications-Anchorage School District Art Contest
Proofreader: Ms. Kim Kubie

Limited First Edition
Includes Bibliography, Index
ISBN: 0971719209

Page 1
The Flag waves proudly from the NOAA ship Rainier as it approaches the city of Petersburg.
Courtesy of Wyndham Images

Page 2-3
Lituya Bay, a place of unusual beauty, is perhaps an accident of Nature. Mt. Crillion, 12,726 feet high, towers above the other peaks that surround the bay, and Cenotaph Island marks the center. The calm surface of the bay belies its violent creation, a result of its relationship to the Fairweather fault. The calm waters of the bay have unexpectedly taken the lives of unsuspecting sailors, giving rise to the island's name (Cenotaph means empty grave), and seismic forces probably caused the giant waves that denuded the shoreline.
Courtesy of AeroMap U.S.

Table of Contents

Part One

Introduction 6

Chapter One
The Alaskan Legacy 8

Chapter Two
Land of the Midnight Sun 22

Chapter Three
Native Alaskans 38

Chapter Four
A Vibrant Economy 54

Chapter Five
Celebrating Alaska 68

Chapter Six
America's Last Frontier 82

Chapter Seven
The Great Outdoors and the Sporting Life 94

Chapter Eight
North to Alaska 108

Part Two: Prominent Communities of Alaska 128

Part Three: Team Alaska
Networks 148
Natural Resources 176
Business And Professional Services 188
Building Alaska 206
Quality Of Life 226
Marketplace 240
Government And Community Organizations 250

Team Alaska Index 276
Bibliography 280
Index 282

Introduction

Alaskans enjoy a reputation as pioneers. Earned during the gold rush of the late 1890s, this image still rings true more than 100 years later—but in a different sense. Although we think of ourselves as the Last Frontier, we eagerly embrace the adventure of crossing the threshold to a new century. Alaska truly is the Frontier of the Future.

Alaska is rich in oil, timber, fish, coal, and other minerals. Development of our assets while preserving Alaska's beauty not only supplies Alaska and the nation with necessities, but creates jobs and educational opportunities for our residents.

Alaska Natives, and those who followed them into the country, comprise Alaska's multi-cultural population. The diverse cultures are prevalent in many communities and offer rich resources of art, music, dance, and other traditional activities. The education of Alaska's youth is a high priority and cultural diversity enriches our schools.

The spirit of Alaska is one of enterprise and adventure, of independence and pride, of respect for the land, and enduring camaraderie among the pioneers who live and work on what is still known as The Last Frontier. The people who live in Alaska are proud to be called Alaskans.

This is Alaska. Our future is bright! Welcome.

Frank H. Murkowski
Governor of Alaska

Isanotski Peaks, also known as "Ragged Jack," is located on Unimak Island in the Aleutian chain. Isanotski is a stratovolcano that reaches 8,025 feet.

Chapter One

THE ALASKAN LEGACY

In Alaska's colorful past, cycles of boom and bust were the order of the day for early prospectors, pioneers, and businessmen and women. That is no longer the case. The up-and-down cycles have been replaced by steady growth as the state has matured politically, socially, and economically. Solidly based on environmentally sound development of its natural resources, the new Alaska economy is successfully reaching out to embrace the communications, engineering, construction, transportation, and air cargo industries, to name just a few. With highly educated and trained work forces in every field, Alaska's future is very bright and growing ever brighter.

**100 Words by
Peter Leathard**
VECO Corporation

The Alaskan Legacy

Alaska is like a nation within a nation. Few places in North America have witnessed a more fascinating historical timeline, unpredictable economic and political swings, and dramatic natural history. Across 367 million acres, from the early eighteenth century to date, the state has evolved from a vast unknown in the modern world to one of the world's most significant players in economic and political arenas around the globe.

The official discovery of Alaska was by Vitus Bering, a Danish sea captain commissioned by Peter the Great to explore the Northwest coast of Alaska. The British, Spanish, and French were also exploring the coast of Alaska. Prior to this discovery, various Native peoples populated Alaska. Eskimos lived along the far north coasts and islands of the Bering and Arctic Seas, Cook Inlet, and Kodiak Island. While Eskimos can be divided into more than 20 groupings, linguistically they fall into three main groups: the Inupiat and Yup'ik primarily in the Far North region and the Sugiaq to the south. The Aleuts (known to themselves as the Unagon, or "original people," before the Russians called them the Aleuts, a name they quickly adopted) laid claim to the Alaska Peninsula and the Aleutian Islands as well as Kodiak Island—which they shared with the Alutiiq. The Athabascans, a largely nomadic people with linguistic ties to the Apache and Navajo, inhabited the Interior. Along the Southeast section of Alaska lived the Northwest Coast Cultures; specifically, the Tlingit, who dominated the region, and the Haida, who settled on Prince of Wales Island from Canada.

Written historical records do not exist until 1741, when Bering began to explore Alaska's Aleutian Islands, the coast bordering the Gulf of Alaska, and the Southeast panhandle. After establishing Russia's claim to the Northwest, Bering died from scurvy on an island named after him—Bering Island. Before contact with Russians and Europeans, the history of Native Alaskans was preserved through the oral tradition.

Unloading fish onto a boat, circa 1917.

A vintage snowplow train is on display by the Historic Potter Section House, which now serves as headquarters for Chugach State Park.

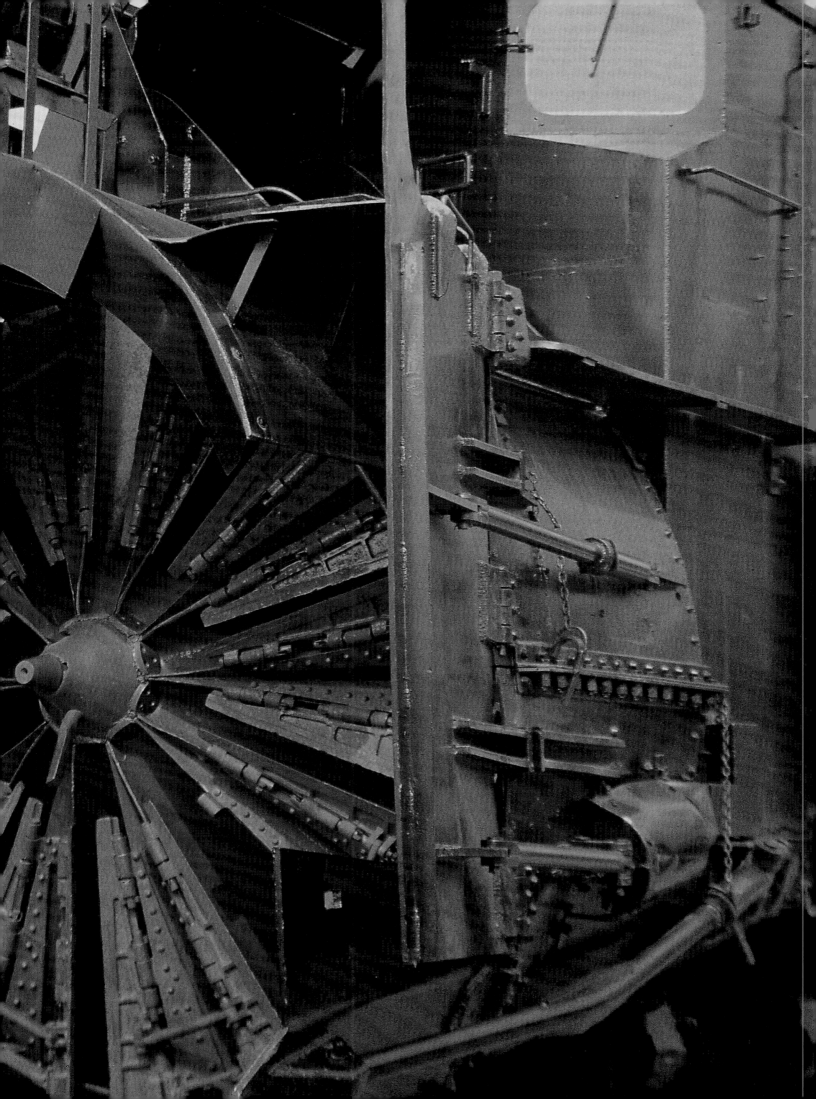

William H. Seward.

"The Magnet" roadhouse near Bonanza was a boon to weary travelers.

European Discovery

The Europeans brought with them the sweeping changes of modern civilization, some of which seriously impacted the Native people and their traditional cultures: international power struggles, alien cultural influences, capitalistic greed and discrimination aided and enforced by modern machinery, and Western disease. Beginning in 1745, Russian fur traders wintered on the Aleutian Islands and along the Alaska Peninsula, establishing settlements on Unalaska in about 1770. In 1784, a Russian fur company headed by Gregory Shelikoff made the first permanent settlement in Alaska at Three Saints Bay, Kodiak Island. Unregulated exploitation of the fur resources by rival companies severely depleted the resource and led to the killing and enslavement of the Aleut Natives, until the Russian American Company was chartered in 1799 and given control over all of the Russian territory in America. Under the rule of its first manager, Alexander Baranov, there was a period of about 20 years when the fur harvesting was orderly and systematic.

Signing the Treaty of Cessation 1867.

Russian-Tlingit War

Tlingits lived alone on Baranof Island until the Russians arrived in 1799 and established a fort north of what is now Sitka. The Tlingits attacked the fur-trading outpost in 1802, killing nearly all of the Russians and their Aleut slaves. The Russians had to resort to a combined naval-military operation to drive the Natives out in 1804. In 1805, the Tlingits killed all of the Russian residents that were living at present-day Yakutat. The cost of administering the Alaska territory started to drain the Russian homeland and this, along with the disappearance of the sea otter and fur trade, brought about the end of the Russian period. Two years later, Alexander Baranof, the manager of the Russian-American Company, attacked with the cannons aboard the *Neva*, and after a six-day battle the Tlingits slipped away. The Russians renamed the settlement New Archangel Bay, which was later renamed Sitka, meaning "by the sea" in the Tlingit language.

Nome in 1900.

It was a long and bitter battle to get Congress to approve the purchase and then to appropriate the money for it. Seward prevailed in completing the purchase, but he became the butt of popular jokes over it, with the newspapers declaring the purchase to be "Seward's Folly" and Alaska to be "Seward's Icebox." When asked what was the most significant act of his career, he declared, "The purchase of Alaska! But it will take a generation to find that out."

Seward's Day, an Alaska state holiday, commemorates the signing of the treaty on March 30, 1867, finalizing the purchase of Alaska by the United States from Russia. On the misty afternoon of October 18, 1867, at the City of Sitka, on the chilly Alaska coast amongst the firing of Russian and American cannons, the Imperial Russian flag came down over Russian America. The Stars and Stripes were raised up the 90-foot flagpole. "Seward's Icebox" had become a part of the United States.

Alaska Purchase

After the end of the fur trade and a number of political factors, Russia sold Alaska to the United States in 1867. President Andrew Johnson's Secretary of State, William H. Seward, was responsible for negotiating the purchase of Alaska from Russia. Seward was so adamant on purchasing Alaska that he started negotiating with Russia before he had authorization from the President. His original offer was for about $5 million. While the Russian Minister was taking the offer to the Czar, Seward asked the cabinet for authority to offer $7 million, and much to his surprise, his request was approved.

By March 23rd, both parties had reached an agreement on the main points of the purchase, including the $7 million figure that had been reached, and asked for authority to sign the treaty. Seward wanted so much for the treaty to be signed that he opened the State Department that evening after hours and made the Russian delegation welcome. The Russian minister wanted to clarify some of the smaller points before signing the agreement. Seward refused to consider them, but in turn added another $200,000 to the final purchase price. This translated into approximately 2.5 cents per acre for 586,400 square miles of territory— more than twice the size of Texas.

Statehood

The Alaska "district" was administered for the first ten years of its existence by the Army. Following that, the Treasury Department did so for two years and the Navy for five years. During the entire time, there was no code of law. Alaska's first governor was appointed in 1884, and a legal code was finally put in place between 1898 and 1900.

The capital was moved to Juneau from Sitka in 1906, and the first official delegate represented the state at the National Congress that same year. Alaska finally became a territory in 1912. By the time it became the forty-ninth state on January 3, 1959, Alaska's history read like a well-crafted novel, woven together with compelling tales of good guys and bad guys, romance, and page-turner adventures that keep each reader on the edge until the very end.

Chapter One: The Alaskan Legacy

The Pioneer Mining Company in Nome displays one week's output of gold, circa 1900.

The trek of those in search of gold was grueling. These would-be miners set about crossing Chilkoot Pass.

Skagway train in the center of town.

THE GOLD RUSH ERA

In spite of the almost total lack of government, some industries were maintained and even added in the early years of American administration. First, fish salteries and later fish canneries were established, most of which operated in the summer only. Although the Nome Gold Rush would not start for another 20 years, in 1876 gold was discovered south of Juneau at Windham Bay and Juneau was established in 1880 following a gold find there. In 1896 the discovery of gold in the Yukon Territory of Canada brought thousands of gold seekers through Alaska on the way to find their fortunes in the gold fields. Another strike was found in Nome, along with several in the Interior along the Yukon River. The last major discovery brought Fairbanks into being in 1902.

The gold rush brought thousands of people to Alaska, both the gold seekers and the people who provided them with supplies and services. Before the gold rush, it was estimated that there were only about 32,000 people in Alaska, and of those about 80 percent were Native Alaskans. By 1900, there were over 63,000 people, an increase of about 26,000 non-Natives. Native Alaskans had survived and thrived for thousands of years by living on the fish, wild game, moose, caribou, berries, and roots that the land provided. The gold rush brought people to parts of Alaska where white men had rarely gone and matters of survival were critical. The miners and the new towns were a threat to some Alaska Natives' semi-nomadic lifestyle and traditional dependence on the land.

In just a few decades, towns, transportation, and communications systems resulted from the development of mining, and shipment needs led to the establishment of several ports along the Gulf of Alaska and Prince William Sound. Cordova, Valdez, and Seward were established and other, smaller towns developed as a result of a growing fishing industry. Railway lines improved the efficiency of moving goods around the state. Schools were established and rustic trails were turned into roadways. Population centers began to require better governance. While prospecting for gold continued, at the same time, an interest in harvesting other valuable minerals and resources expanded.

The White Pass & Yukon Railway's rotary snowplow clears the line.

Klondike Gold Rush

The Klondike gold strike in the Yukon Territory was one of the last gold rushes in which prospectors could hope to dig out a fortune from the earth. Because it came so late in time compared to other major gold strikes, and because some miners did take home millions in spite of the difficult environment, this gold rush left a lasting mark on the American imagination.

The Klondike Gold Rush was significant not only because it was the last great gold rush, but also because it increased awareness of the northern frontiers of Alaska and Canada. Unimpressed, the press had labeled the purchase of Alaska as "Seward's Folly" or "Seward's Ice box." Alaska and the Canadian Northwest, including the Yukon Territory, remained sparsely populated until the end of the century. When the U.S. Census Bureau declared the western frontier closed in 1890, interest in Alaska grew. While there still were millions of acres of empty space in the lower states and territories, more people began to venture north, toward the lands they recognized as the Last Frontier. The discovery of gold, first in Yukon Territory and then in Nome, Alaska, raised the public's interest in what the Far North had to offer.

Early Economic Development

Many changes took place in the Yukon as a result of the gold rush. A railway was built from Skagway, Alaska, to Whitehorse, Yukon Territory, in 1900. The population of Whitehorse swelled to 30,000 the same year. The gold-bearing gravel found between the Yukon and Klondike Rivers brought as much as $22 million in 1900, but it fell to $5.6 million by 1910, when most of the stampeders had left for Alaska, returned to Seattle, or set out to other regions.

The eighteenth century introduced cultural and economic influences that would shape Alaska for years to come. In the 100 or so years between the middle of the eighteenth century when Russian explorers and fur traders recognized the opportunity that Alaska presented and the early 1900s, the territory witnessed exploration and discovery by Europeans who were quick to capitalize on the pelts of fur seals and otters. The dwindling of the fur trade and involvement in the Crimean War led Russia to sell Alaska to the United States in 1867. By the end of that century, oil, coal, and fish were also being recognized for their value, and the first gold claims staked in Cook Inlet were feeding the gold fever that had by now spread to the Lower 48 states and was drawing hardy, sourdough spirits from across the country.

Fish canneries, such as this Hoonah cannery in 1917, were an important part of Alaska's early economic development that took place after the gold rush.

Chapter One: The Alaskan Legacy

Several men manage their canoes at the mouth of the Chilkat River, near Yun-day-stuck-e-yah Village, also known as Gantegastaki, circa 1895.

Canoeing on the Yukon, 1906.

An old sourdough tries his luck at panning for gold.

"Just Who're You Callin' a Sourdough?"

There may be a bit of confusion among the uninitiated regarding the term "sourdough." The reader may stumble a bit when realizing that the text is referring to a person, *not unbaked bread. A "sourdough" is a nickname for a miner and the nickname is, indeed, a reference to unbaked bread.*

Miners that headed out to seek their fortunes packed all they needed with them, and that included all their food. Yeast did not last long enough and chemical leavenings such as baking soda or alum were unreliable and susceptible to water damage. So, the innovative "sourdoughs" relied on an ancient way of preparing bread. Each time they baked bread, they kneaded the old starter with new flour and water and let it rise. Before baking, they set aside a new piece for the next loaf. Since the starter was delicate and needed to be kept warm, it was often kept in a leather pouch on a belt, next to the body.

The thing that brought the "sour" to the dough came from San Francisco. What was unique to San Francisco sourdough was Saccharomyces exiguus *yeast, which did not favor maltose sugar and left it intact in the bread, thereby lending it its distinctive sour taste. The bread was popularized during the gold rushes in California, and the starters spread all the way to the Yukon. And so "sourdough the bread" and "sourdough the miner" became fixtures in western history and lore.*

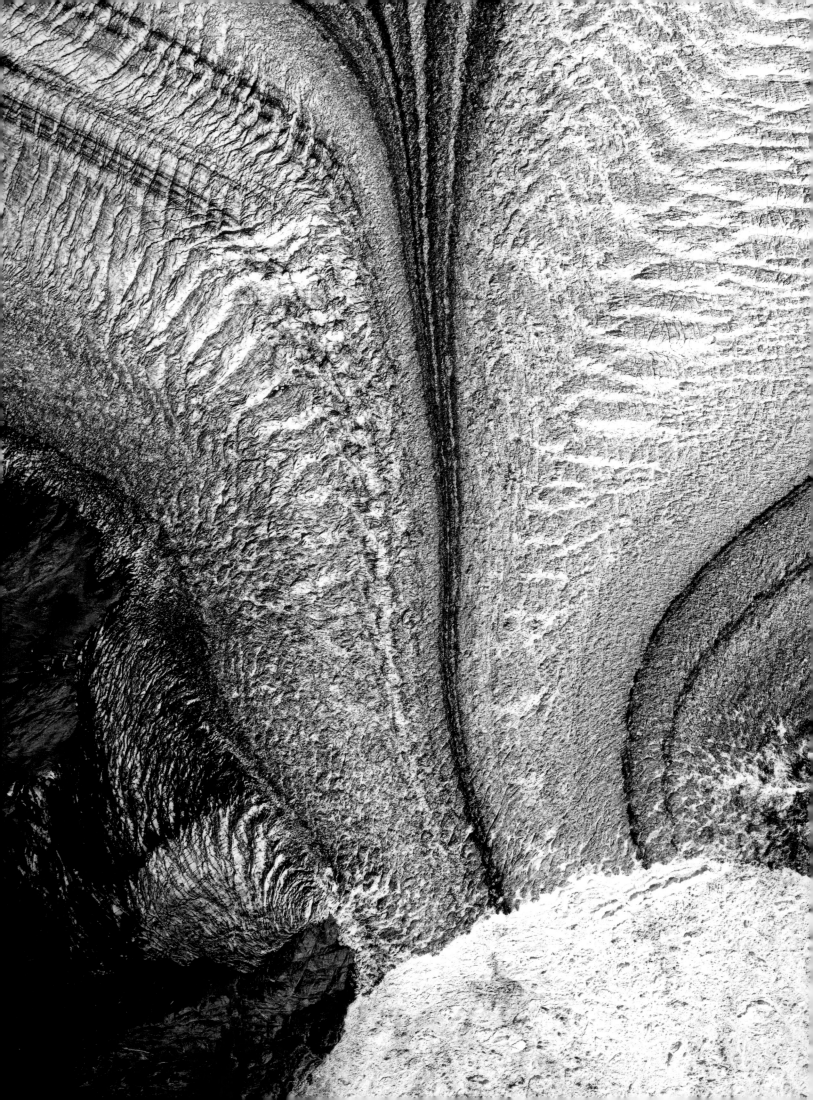

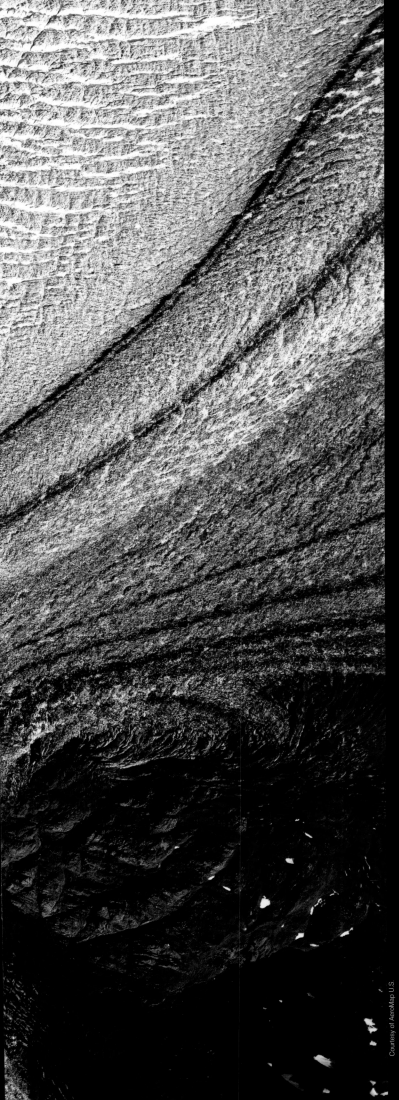

Chapter Two

LAND OF THE MIDNIGHT SUN

There is a rhythm to Alaska, first heard eons ago on distant walrus-skin drums. Early Russian, Spanish, and English explorers discovered diverse Native cultures throughout the land and joined in, eventually harmonizing their own rich and colorful influences into the heartbeat of the Great Land. This blending is manifested at places such as Kotzebue, Sitka, Valdez, and Cook Inlet and on the varied faces of Alaskans from Barrow to Metlakatla. With the discovery of gold at Nome and Fairbanks, the rhythm reached fever pitch, thereafter followed cycles of boom and bust—but the heartbeat goes on toward a sustainable future.

100 Words by
Jeff Staser
Denali Commission

Columbia Glacier.

Land of the Midnight Sun

From the Tongass National Forest that blankets the Southeast panhandle, to the 80,000 acres of tundra that defines the North Slope, and the rich fishing waters that link coastal communities of the Southwest region, Alaska defies one simple description. Great rivers that once forged this broad lowland between the Alaska Range and the Brooks Range mark the Interior. Southcentral has some of the state's largest cities, and just minutes away from these urban areas is one of Alaska's few farmland and agricultural industries. It is this multi-faceted nature that feeds the love affair that 627,000 residents have with this land they know as home.

The diversity does not stop there. Within each of the larger five geographic areas of the state there are still smaller sub-regions that can often be substantially different from their neighboring sub-region. Like states within a state, they have their own unique topographies, economies and cultural influences, though in many cases only a short plane trip separates them. Consider Southwest Alaska. Within this one region, the panorama ranges from the gentle grasslands of the Aleutian Islands to the surreal volcanic landscape of Katmai National Park. Kodiak, one of the larger communities in the region and the second largest island in the United States after Hawaii, is steeped in Russian history, while the neighboring Bristol Bay area is home to Yu'pik and Athabascans. Unalaska is perched on the Outer Continental Shelf in the Bering Sea, an underwater nursery providing some of the world's richest habitats for several species of commercial fish and marine wildlife. All this is in an area no larger than the size of most states.

Southeast

Alaska's "banana belt" stretches 500 miles from Icy Bay, northwest of Yakutat, to the Dixon Entrance at the United States-Canada border, and is accessible by air and water only. Although the climate is mild and seldom drops below freezing, Southeast communities are known for being among the wettest in the world, with an average rainfall of anywhere from 80 to 200 inches in the various communities and 400 inches in higher mountain elevations.

Sea lions take a flying leap into the sea.

Kotzebue, Alaska.

Outsiders often learn about this area because of the Inside Passage's fame. The Inside Passage is the maritime highway that runs between Bellingham, Washington, and Haines, Alaska. The waters of the Passage are sailed by Alaska's statewide ferry system and by the largest cruise ship lines in the world. The locals have a different appreciation for the mystic seaside communities, due to their familiarity with the Tlingit, Haida, and Tsimshian cultures, and their regard for the marine wildlife that is simply—unbelievable.

The 17-million-acre Tongass National Forest that covers most of the panhandle is the largest standing old-growth forest in North America. In addition to providing forest products for national and international markets, the Tongass is home to some of Alaska's most magnificent wildlife, including bald eagles, grizzly bears, five species of Pacific salmon, and smaller mammals and waterfowl.

Massive ice fields; glacier-scoured peaks; steep, riven valleys; craggy, jagged fjords; more than 1,000 named islands, and countless unnamed islets and reefs characterize this vertical world where few flat expanses break the steep lines.

Anchorage Opera.

SOUTHCENTRAL

People are the defining factor of this region. About two-thirds of the state's residents live in the communities of Anchorage, the Kenai Peninsula, Prince William Sound, and the Matanuska-Susitna Valley. Anchorage alone is home to over half the state's 627,000 residents, and the neighboring Matanuska-Susitna Valley has the highest population growth in Alaska and is poised to be the state's leading population magnet in the next 10 years. Available, affordable land with prime views and a rural lifestyle make this area attractive to families despite the 45-minute commute to Anchorage.

Where there are people, there are the other predictable signs of population centers. Due to its global position in relation to international flight routes, Anchorage is often referred to as the "Air Crossroads of the World." Anchorage is the gateway for

The Evangeline Atwood Concert Hall, the largest venue at the Alaska Center of the Performing Arts in Anchorage, seats approximately 2,000.

international travelers and trade, as well as the commerce and communications hub, and the jumping off point for all visitors traveling to other parts of the state. Oil companies, the state's leading banks, government offices, and headquarters for large non-profits organizations and Native corporations are based there, along with the Ted Stevens Anchorage International Airport, the Alaska Railroad, and the Port of Anchorage, which are the largest gateways in the state.

The Southcentral region has a milder climate than the Interior with average temperatures of 15F degrees in January and 58F degrees in July, and an average snowfall of about 70 inches a year, with about 20 hours of summer daylight and a minimum of five hours in the winter. Residents of Southcentral also enjoy having the advantages of living in or near an urban environment, just minutes away from many of Alaska's most pristine playgrounds. The 5.6-million-acre Chugach National Forest—the second-largest in the country after the Tongass—runs south and east of Anchorage along the coast bordering this region. Despite its size, a majority of the Chugach is accessible by foot, boat, airplane, or via one of Alaska's designated Scenic Byways, the Seward Highway. The 3,500 miles of spectacular coastline along Prince William Sound, home to Valdez, Cordova, and Whittier, is a paradise for outdoor adventure in the forms of kayaking, boating, fishing, and wildlife viewing. The Kenai Peninsula, known worldwide for its "combat" fishing, is one of the few places where an angler can land a 50-pound king salmon. The Kenai is also a place where Alaska families may find steady employment with the oil companies or the chance to earn long-term livelihoods through Alaska-owned businesses, especially in the visitor industry so important to this region.

Southwest

This dynamic area is divided into five diverse sub-regions, each with a distinct personality shaped by historical, cultural, and economic developments that not only define this region, but also influenced the development of the entire state. Kodiak Island, the Aleutians, Bristol Bay, the Pribilof Islands, and the Alaska Peninsula comprise this predominantly coastal region. Bethel, Dillingham, Kodiak, King Salmon, Naknek, St. Paul, and Unalaska are the region's population centers.

Southwest Alaska covers an area of more than 62,000 square miles. It includes more than 50 communities and has a combined population of just over 30,000 people. The waters surrounding the region are some of the most productive in the world and support a variety of commercial fisheries including salmon, herring, halibut, Pacific cod, sablefish, pollock, Atka mackerel, arrowtooth flounder, rockfish, crab, and other shellfish. These fisheries are the foundation of the region's seafood industry and provide an economic base for many of the area's communities.

Everywhere one looks, the sea is part of what makes this region so mystifying. Windswept islands and active volcanoes are the thumbprint of the Aleutian Islands. Throughout the entire region, fishing grounds are also home to sea mammals, and the surrounding coastline provides critical habitat for grizzly and black bears, elk,

Whittier, a small community that was of great service to the military in World War II, was recently opened to automobile traffic. Once only accessible by boat or by train, a tunnel leading to Whittier was recently enlarged to accommodate both trains and automobiles, thereby opening the small town to a regular flow of visitors by car and bus. For the tiny hamlet on the coast of Prince William Sound, the possibilities brought about by these changes may be endless.

Chapter Two: Land Of The Midnight Sun

Pavlof Volcano is part of the Aleutian Range, near the southwest tip on the Alaska Peninsula. Approximately 580 miles southwest of Anchorage, Pavlof is 8,250 feet high, making it one of the tallest volcanoes in Alaska. It is also one of the most consistently active.

Chapter Two: Land Of The Midnight Sun

Ketchikan, Alaska.

Right: Chief Shakes Tribal House on Chief Shakes Island near Wrangell is on the National Register as a historic monument.

mountain goats, and millions of seabirds that migrate to the region from around the globe. Native Alaskan art and culture reflect their long relationship with the water in the materials from the sea and style of work that artisans use.

Southwest Alaska has a rich historic and cultural heritage. The region is home to four Native Alaska groups: Aleut, Yup'ik, Athabascan and Alutiiq, and it was once an important part of Russian America. The region also includes a number of former military sites that were of strategic importance during World War II and the Cold War.

While most residents are still linked to the commercial fishing industry, tourism is an economic-diversification strategy for a number of communities in this region. More and more small businesses are offering guided sportfishing, eco-tours, and the other services needed to host visitors, such as restaurants, gift shops, and taxi services. Visitors are drawn to Southwest for its diverse, scenic beauty and the opportunity to explore off the beaten path.

The Aleutian and Pribilof Island climate is cool and often unforgiving. Summer temperatures range in the 50s and winter readings are in the 20s and lower, with up to 80 inches of precipitation annually. Constant, year-round winds have left most of the area treeless, and fog can settle for days.

Kodiak Island and the six villages on the Island are warmer than the rest of the region, but are known for having some of the wettest and most unpredictable weather in the state with 80 inches of rain a year and an average temperature of 40 degrees.

INTERIOR

This is a land of superlatives: temperature extremes from hottest to coldest; the highest peaks and broad, flat expanses that go on for miles; and awesome, serpentine rivers that carve their way through the Interior and across the state, including the largest in Alaska, the mighty Yukon.

The 215-foot NOAA ship Miller Freeman *crests an 80-to-90-foot wave in the Bering Sea.*

This central part of the state, between the Alaska and Brooks Ranges, has winter temperatures that commonly drop to –50F degrees, bringing with them ice fog that hovers so low, it is blinding. Conversely, this is the only place in the state that has summer temperatures that are typically in the 90s. The semi-arid climate means that while the temperatures vary greatly, there is only about 12 inches of precipitation per year. The Interior region is also known for its intense, frequent northern lights during winter months that attract Japanese visitors who believe they enhance fertility. Mt. McKinley—the highest peak in North America at 20,320 feet, and the most popular visitor attraction in Alaska—is the crown jewel of Denali National Park, and can be seen from Anchorage, which is 250 miles away.

Sprawling forests of birch and aspen add vibrant green and gold to the landscape. Spruce blankets many of the slopes, and cottonwoods thrive near river lowlands. In the northern and western reaches of the Interior, the North American taiga gives way to tundra. In the highlands above tree line and in marshy lowlands, grasses and shrubs replace trees.

Fairbanks, the second largest city in Alaska after Anchorage, is in the heart of the Interior and is also the transportation, communication, and business hub for all points north, including the North Slope. Healy, just southwest of Fairbanks, has the state's only operating coal mine and produces coal to generate electricity for the area, including Ft. Richardson and the University of Alaska. Delta Junction, 100 miles east of Fairbanks, is where Alaska's first large-scale agricultural project was launched in the early 1970s, resulting in the development of more than 80,000 acres of land. Delta-area farmers continue to plan for the future expansion of the industry. Finished beef, dairy beef, and swine are strong possibilities for development, as well as bison, elk, European boar, and reindeer ranching. The rest of the Interior relies primarily on a subsistence economy, sometimes combined with a cash economy where fishing or seasonal government jobs are available.

Alaskaland in Fairbanks offers a variety of sights and amusements that give the visitor a feel for frontier life. Full-sized mining exhibits and a real paddleboat turn back the hands of time.

Chapter Two: Land Of The Midnight Sun

Chapter Two: Land Of The Midnight Sun

Downtown Juneau.

These ladies are waiting to serve up the next tasty beverage at the Red Dog Saloon in Juneau.

Far North

Located just above the Arctic Circle, the Far North region is as challenging as it is beautiful. The 88,000 square miles of open tundra are laced with meandering rivers and countless ponds. Short, cool summers with temperatures between 30F and 40F degrees allow the permanently frozen soil to thaw only a few inches. Winter temperatures are well below zero, though the Arctic Ocean moderates temperatures in coastal areas. Severe winds sweep along the coast and through mountain passes. Most areas receive less than 10 inches of precipitation a year, but the terrain is wet in the summer because there is little evaporation and the ground remains frozen.

The Far North region, interchangeably referred to as the Arctic, has traditionally been the home of Inupiat Eskimos. Barrow is the northernmost settlement in the nation and the largest Eskimo community in the world. As the seat of the 88,000-square-mile North Slope Borough, Barrow rightfully earns the designation of the world's largest municipality. Eight other Inupiat subsistence villages dot the land, where petroleum-related jobs support most of the region's residents and subsistence hunting and fishing fill any economic holes left by the oil industry.

Just 200 miles east of Barrow is Prudhoe Bay, the largest oil field in North America. Prudhoe is at mile marker 0 of the 800-mile Trans-Alaska Pipeline, which snakes across the state to Valdez where oil is loaded onto tankers to be exported worldwide. Prudhoe Bay is a prime example of how modern technology has been adapted to co-exist with a unique environment.

In the central Brooks Range, Anaktuvuk Pass lies on a historic caribou migration route. Vast herds migrate through this area. There are also millions of acres of America's finest wilderness park lands at Cape Krusenstern National Monument, Kobuk Valley National Park, Noatak National Preserve, Selawik National Wildlife Refuge, and the Gates of the Arctic National Park and Preserve. Currently, these areas attract only the very experienced outdoors enthusiast, and many willing to brave this area's wilderness rely on professional guides to escort them on their journeys.

A trademark scenic view of the Southeast region is the lovely treed, steep hillside with the tumbling splash of a waterfall.

Alaska and Its Amazing Glaciers

Alaska is renowned for its glaciers. Glaciers cover over 29,000 square miles—about five percent of the entire state! Owing to its unique geology and northern latitude, Alaska easily has the most glaciers of any state in the nation.

People most often think of glaciers as being giant "rivers of ice." In fact, glaciers are categorized by size, including ice fields, ice caps, and ice sheets; and also categorized by location, including alpine, valley, and piedmont glaciers. Glaciers form from a long-term accumulation of ice, snow, water, rock, and sediment. Gravity puts the ice in motion. Since glaciers form mostly in Polar Regions and at high altitudes, they are not common sights for most people. So it is perhaps surprising to realize that glacier ice covers approximately 10 percent of the Earth's land area and about 80 percent of the planet's fresh water is frozen in ice sheets and glaciers.

When traveling through Alaska, particularly when taking a cruise or a flight-seeing tour by helicopter, the adventurer will undoubtedly encounter, perhaps even walk upon, a valley or piedmont glacier. These vast "rivers of ice" flow from ice caps or high mountain basins and carve the landscape on their way down. To experience the phenomenon of a glacier is drama itself. To stand upon a glacier and sense its awesome power can be a humbling experience; and to watch from a boat—from a safe distance of course—as a glacier calves (a piece of the glacier breaks free) is nothing less than thrilling.

Perhaps most memorable of all is that special, otherworldly blue. It is best seen just after calving, or deep down a crack or crevasse. It is glacier blue. Glacial ice often appears blue when it has become very dense. Years of pressure force out tiny air pockets between crystals. If the ice is white, many air pockets exist and all light is reflected back. But when the ice is blue, the air pockets are gone and the ice has absorbed most of the colors of the spectrum except the blue, to the good fortune of all those who see it.

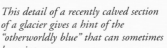

This detail of a recently calved section of a glacier gives a hint of the "otherworldly blue" that can sometimes be so intense.

Chapter Two: Land Of The Midnight Sun

According to the Geophysical Institute, the Great Gorge at Mt. McKinley may be the deepest gorge in North America, and possibly the world. The Great Gorge was found to be nearly 9,000 feet deep—deeper than the Grand Canyon or the Valleys of Yosemite National Park. Forty-mile-long Ruth Glacier fills the bottom of the gorge.

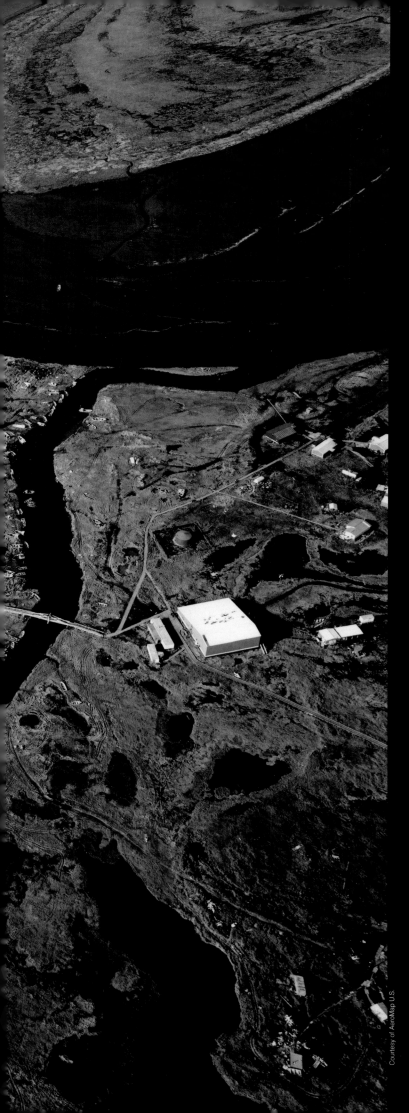

Chapter Three

NATIVE ALASKANS

I feel blessed to have been born in Alaska, raised by traditional Tlingit grandparents. For Tlingits, respect for every living thing provides the foundation for "life in balance." Our very existence derived from the rich resources of our homeland. Enactment of ANCSA changed life dramatically. A new view was introduced, challenging our subsistence lifestyle. Our traditional trade-and-bartering system needed to coexist with an alien concept of corporate culture. It had immense economic, political, and social impact. Our challenge as Alaska Natives is to move within this new world, retaining traditions that enabled us to survive millennia as strong, resourceful people.

100 Words by
Ethel Lund
(Tlingit Name, Aan wugéex), Tlingit Elder

Kipnuk is on the west bank of the Kugkaktlik River in the Yukon-Kushokwim Delta. The community is four miles inland from the Bering Sea. Yup'ik Eskimos have inhabited the region for thousands of years.

Native Alaskans

As varied as the topography and weather in each region, so is the Alaskan culture. Native Alaskans account for 99,000 (2001 Census), or about 16 percent, of the state's population. There are many ways in which anthropologists, sociologists, the government, the general population, and the Native peoples themselves divide and categorize the cultures and sub-cultures of the Native peoples of Alaska. In the most general terms, there are four distinct Native cultural groupings in Alaska, and dozens of sub-cultures. The primary Native cultures are the Eskimos of the Far North and the Southwest, which include the Inupiat, the Yup'ik, and the Sugiaq; the Aleuts of the Aleutian Islands; the Athabascans of the Interior; and the Northwest Coast Cultures, which include the Haida, the Tlingit, and the Tsimshians of the Inside Passage.

These groups have distinct cultural traditions, art forms, and languages. Many Native people are bilingual in English and their own dialect. Although many live in rural villages throughout the state and along the rivers and coastline, about 23,000 live in larger communities of Anchorage and Fairbanks.

While the theory is being challenged at present, it is generally believed that Native Alaskans are the descendants of nomadic explorers who crossed a prehistoric land bridge that linked Siberia and North America. Those early travelers became the ancestors of Alaska Natives and American Native nations in the Lower 48. Living a nomadic lifestyle, early Alaskans adapted well to a unique, often harsh environment and developed a spiritual world rich in tradition. When Europeans first came in contact with Native Alaskans in 1741, Native Alaskans lived within separate, well-defined regions, and the different ethnic groups seldom mixed, with the exception of contact for purposes of trade. Today, Native Alaskans play a vital role in the state's economy and society, while preserving their traditional cultures.

In Southeast, the herring, salmon, deer, and other plentiful resources allowed the Haida and Tlingit to settle in permanent villages and develop a culture rich in art. The Athabascans lived mostly in the

Bald eagle.

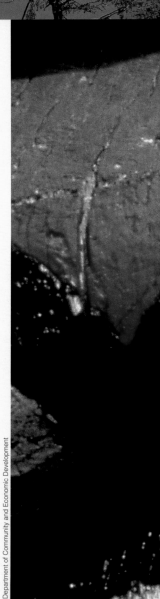

Totem face.

Chapter Three: Native Alaskans

Aleuts in Baidarkas, 1826.

Athapascan village of Taral on the Copper River, 1885.

Interior and migrated from one seasonal subsistence camp to another to take advantage of seasonal fish, waterfowl, caribou, and other game. The coastal Eskimo and Aleut subsisted primarily on the rich resources of the rivers and the sea.

Alaska's Tsimshians moved in 1887 from their former home in British Columbia to Annette Island in Southeast Alaska. Today, about 1,200 Tsimshian now live in Metlakatla, and like many Southeast residents, are fishermen. The Haida excel in the art of totem carving and are noted for skilled working of wood, bone, shell, stone, and silver. About 1,800 Haida live in Alaska, 300 of whom live in Hydaburg on the south end of Prince of Wales Island. It is believed they migrated from Canada.

The Tlingit migrated west from what is now Canada 10,000 years before the first European contact and commercially dominated the interior Canadian Natives, trading eulachon oil, shells, and smoked seafood products for furs, beaded clothing, lichens, and other products unavailable on the coast. They are also master totem pole carvers. Many totem poles provide a record of major events in the history of a family or clan.

The Interior region, directly in the middle of the state, is home to almost 14,000 Athabascans, nomadic tribes that hunt for caribou, moose and fish. The Eskimo traditionally lived in villages along the harsh Bering Sea and Arctic Ocean coastlines, where their houses were barabaras, dwellings built partially underground and covered with sod.

The Aleut have traditionally lived on the Alaska Peninsula and along the Aleutian Island Chain. When the Russians reached the Aleutians in the 1740s, practically every island was inhabited. Decimated by contact with the white man, only a few Aleut settlements remain, including the Pribilof Islands of St. George and St. Paul, where Native Alaskans handle fur seal herds for the federal government. The Aleut lived in permanent villages taking advantage of the sea for food. Their original dwellings were large, communal structures, housing as many as 40 families. After the Russian occupation, they lived in smaller houses, many adopting the Russian-style log cabins. Today, many Aleuts are commercial fishermen.

Tlingit camp during the fishing season at Port des Francais (now called Lituya Bay), circa 1785.

Chapter Three: Native Alaskans

This Eskimo oomiak, with a mast for a square sail, is made of split walrus hide.

ANCSA

In the spring of 1867, the sale of Alaska from Russia to the United States was completed. One brief reference in the treaty addressed the issue of Native Alaskans' status, rights, or land ownership: "The uncivilized tribes will be subject to such laws and regulations as the United States may, from time to time, adopt in regard to aboriginal tribes in that country." The issue would not be dealt with again until the passing of the Alaska Statehood Act in 1958 when this new legislation—while acknowledging the right of Natives to lands they had inhabited for thousands of years—authorized the state government to use 103 million acres as it saw fit.

Native Alaskans saw the land as the key to enabling them to actively participate in the new economic development of the state. Certain factors worked in their favor. First, there was a push to develop land for oil; second, the state was anxious to get land promised to it under the Statehood Act; and third, the Alaska Natives felt it was imperative to value and save their land as well as prosper in a changing modern world. All these elements worked as catalysts to complete one of the most complicated deals ever made—everyone had to work together to get what they wanted. Working with the White House and Congress, the Alaska Federation of Natives came up with a resolution. On December 18, 1971, President Nixon signed the Alaska Native Claims Settlement Act (ANCSA) into law. Regarded as one of the most significant land transactions and complex acts in history, after four years of heated debate, ANCSA gave Native Alaskans $962.5 million and 44 million acres of land as compensation for the loss of lands historically occupied or used by their people. Today, lawyers and Native Alaskan leaders are still arguing over interpretations of some aspects of ANCSA.

There were nearly 80,000 Native Alaskans who could participate in ANCSA when it was passed. Most of those affected by the act were in Alaska, but about 20,000 people lived in the Lower 48 state and other parts of the world. "Native" was defined as a citizen of the United States with one-fourth degree or more Indian, Aleut, or Eskimo ancestry, including Natives who had been adopted by one or more non-Native parents. Later, amendments were passed to allow Native Corporations to issue stock to those born after December 18, 1971.

Thirteen regional corporations were formed, including 12 in Alaska and one that was created later to represent Alaska Natives living outside the state. The size of the regional corporations ranged from Ahtna, Inc., with about 1,000 shareholders, to Sealaska Corporation, with about 16,000 shareholders. Others included The Aleut Corporation; Arctic Slope Regional Corporation; Bering Straits Native Corporation; Bristol Bay Native Corporation; Calista Corporation; Chugach Alaska Corporation; Cook Inlet Region, Inc.; Doyon Ltd.; Koniag, Inc.; NANA Regional Corporation, Inc.; and the Thirteenth Regional Corporation. Over 200 village corporations were also created. Most Alaska Natives are shareholders in both a village and regional corporation.

Three Tlingits in dance regalia at the 1904 Potlatch in Sitka.

Chapter Three: Native Alaskans

Both spiritual values and physical environment are strongly represented in Native Alaskan art.

Chapter Three: Native Alaskans

Art

The different types of artwork Native Alaskans create is traditionally determined by the region in which they live and the natural resources available to them. Traditional arts and crafts were produced for ceremonial and utilitarian reasons. These works were not thought of as art in the Western sense, but as object and designs to fulfill specific functions. Native art also reflects spiritual values and the physical environment each group inhabits. Roots, bark, grasses, wood, fur, skins, feathers, and the sea's resources are still used to produce containers, clothing, hunting tools, and ceremonial regalia.

The Inupiat Eskimos of northern Alaska, particularly those along the Bering Sea and Arctic Ocean coastlines, specialize in skin sewing and ivory carving as a result of their subsistence lifestyle which depends heavily on walrus, whales, and seals. Their ivory carving and scrimshaw work is world renowned, and more contemporary work in which stiff baleen is coiled into elegant baskets is also gaining recognition.

The Yup'ik Eskimos of Alaska's Southwest coast are well known for their collector-quality coil baskets created by tightly weaving beach grass. The Yup'ik are also noted for their ivory carving, most of which is done in the winter (leaving the milder seasons to hunt). Yup'ik ceremonial masks, which are carved primarily of driftwood, assembled and painted, are distinctive in the global tribal mask-making tradition.

The Aleuts of the Alaska Peninsula and the Aleutian Chain are famous for their intricately woven basketry. Made from tough, pliable rye grass native to the region, these baskets were once used as a form of currency during the nineteenth century because they were so valued. They are also known for their traditional capes made of sea-mammal gut and their painted bentwood hunting hats and visors, decorated with trade beads and sea-lion whiskers.

The Athabascans of Alaska's Interior have long created beautiful beadwork that is highly prized by collectors of Alaska Native crafts. The traditional use of beads carved of wood, seeds, quills, and shells predates the Athabascan's contact with Europeans who introduced glass trade beads in the mid-nineteenth century. Athabascan art also includes birch-bark baskets.

The Tlingit, Haida, and Tsimshians of Southeast Alaska are known for their colorful totem poles, bentwood boxes, and ornate ceremonial blankets that depict stylized animal forms. Hand-carved silver jewelry and ceremonial masks are also characteristic of these cultures.

Subsistence Lifestyle

State and federal law define subsistence as the "customary and traditional uses" of wild resources for food, clothing, fuel, transportation, construction, art, crafts, sharing, and customary trade." Subsistence uses are central to the customs and traditions of all Native cultural groups in Alaska.

In both rural and urban areas, Native Alaskans statewide live in the modern-day twenty-first century, while still practicing traditional subsistence activities in their communities, revealing a culture that reaches back thousands of years. Those who live off the land, whether by choice or chance, expend considerable amounts of energy in the daunting, sometimes perilous tasks of hunting and gathering. Yet even today, more than 100,000 people in rural Alaska continue to wrestle much of their sustenance from the land and the sea. The knowledge of where to find roots and greens and when to hunt bears and beavers is a part of the Native Alaskan heritage that is passed from generation to generation in all five regions of the state. Most of the wild food harvested by rural families is made up of fish, land mammals, marine mammals, birds, shellfish, and plants. The aquatic species include salmon, halibut, herring, whitefish, seal, sea lion, walrus, and beluga and bowhead whales. Moose, caribou, deer, bear, Dall sheep, mountain goat, and beaver are commonly harvested land mammals, depending on the community and the region.

Whaling

Eskimos believe that the whale gives itself to deserving captains and crews. They honor and respect the whales for providing blubber to heat their homes and food for their people for hundreds of years. The whale is shared by the community, in keeping with Native Alaskans' belief that it is an honor to share.

Subsistence whaling for bowhead is allowed under regulations established by the International Whaling Commission and the Alaska Eskimo Whaling Commission. Irresponsible commercial whaling dating back to the nineteenth century drastically reduced bowhead

Ceremonial canoe ride.

Chapter Three: Native Alaskans

A traditional fish-drying rack in Nome.

Chapter Three: Native Alaskans

The traditional fur-trimmed parka is commonly worn as protection from the bitter cold.

Chapter Three: Native Alaskans

populations, resulting in this species being under international protection. With the cooperation of Alaska's Eskimos, the United States took steps to regulate Eskimo whaling to protect the bowhead without impairing the unique Eskimo culture. The National Marine Fisheries Service and the Alaska Beluga Whale Committee conduct research on the biology, natural history, and traditional knowledge of the Western Alaska population of beluga whales to manage the annual western harvest.

HERITAGE PRESERVATION

Native corporations share the common goal of striving to operate profitable enterprises and be regarded as serious corporate presences, while at the same time preserving their shareholders' culture and heritage in a world that often does not understand or appreciate traditions based on honor, loyalty, family, and tradition. In many communities, modern high schools, housing with heating systems and running water, and full-service grocery stores have contributed to the decline of traditional lifestyles.

While Native Alaskan youth are encouraged to respect their elders' views, often they have different agendas and take little interest in traditional dances and customs, and might choose a McDonald's Big Mac over seal meat. This dramatic change, occurring over a few short years, is almost unbelievable to elders who once hauled ice for water (perhaps some still may) and used dog sleds for travel, while today's youth growing up in bush communities often have little understanding of traditions, and already take modern conveniences for granted. Responding to the wishes of their shareholders, and recognizing that generations of history can disappear in a matter of years, more and more Alaska Native corporations are investing heavily in mentoring, historical preservation, and educational programs so that their youth may develop identities that represent the best of both their heritage and contemporary thinking.

The sunset silhouettes a drying rack and the ribs of an oomiak (a Native canoe).

The Hand-Woven Magic of Qiviut

In villages scattered throughout the tundra and coastal regions, Eskimo crafters carefully and skillfully knit using a very soft, grayish-brown wool-like fiber. It is a very special fiber—one of the rarest, finest, and warmest fibers on earth. Called qiviut (sometimes spelled quiviut and is pronounced kiv-ee-ute), the word for this fiber means "down" or "underwool" in the Eskimo language.

Qiviut is the silky-soft underwool from the Arctic musk oxen that is grown for added warmth during the winter and then shed naturally during the spring months. Eight times warmer than wool, qiviut is not scratchy nor will it shrink in any temperature of water like wool can. Due to its rarity, the price of qiviut has soared to over $175 a pound. Still, demand for the fiber far outstrips the supply. As a result, there is increased interest in domesticating the musk oxen. The question remains, will the quality of the qiviut from the domesticated animal be equal to that of the qiviut from wild musk oxen? It is a question that researchers from the University of Alaska Fairbanks are attempting to answer.

Meanwhile, qiviut continues to be harvested and sent to knitters in Native villages through a co-operative that is owned by approximately 250 Native Alaskan women. The co-operative affords the women the opportunity to bring added income into their families. They knit hats, scarves, stoles, tunics, and nachaqs, which are tubular garments that may be worn around the head like a hood or around the neck in the manner of a scarf or decorative item. The women knit the garments using their village's signature pattern, which may have been derived from anything ranging from beadwork patterns, to artifacts, to nature itself. The resulting garments are deceptively light for the warmth they afford. The grayish-brown color is a pleasing neutral, the knitwork is beautiful, and the fiber is luxuriously soft—enough so to rival cashmere. This unique cottage industry has been successful, with sales via the Internet and through a store in Anchorage.

So from the lumbering musk oxen comes an unexpected beauty. The skillful hands of hundreds of Native women weave the magical fiber called qiviut.

Musk ox.

Chapter Three: Native Alaskans

Qiviut is as soft as cashmere and warmer than wool.

Base Operations Camp on
Barter Island.

Chapter Four

A VIBRANT ECONOMY

As a long-time Alaskan resident, I consider myself fortunate. I've been a successful businessman; Governor; President/CEO of Alaska Railroad; and now Director of Port of Anchorage. My experience, sense of history, and vision compel me to focus on issues I believe vital to our state's growth. Statewide opportunities include the missile defense system, the gas pipeline, Knik Arm Bridge, and ANWAR, among others. These projects will stimulate the state's economy, creating employment opportunities with good incomes and improving the quality of life for Alaskans.

Someone once said, "Build it and they may come; don't build it and they can't come."

100 Words by
Governor William J. Sheffield
Port of Anchorage

A Vibrant Economy

Alaska's economy is based on its natural resources: oil, gas, seafood, scenic beauty, minerals, and timber. However, true economic success rests with Alaskans and a "can-do" spirit that is legendary.

As Alaska's economy continues to move away from boom-and-bust cycles, a more stable picture emerges as typified by a decade of steady growth. Residential and commercial construction, homeownership, and new business start-ups are at an all-time high. Additionally, the state is expected to add jobs at a rate of 1.5 percent annually for the next five years.

With a population projected to grow by 43,000 or 0.6 percent annually through 2012, Alaska faces challenges as it works to strengthen and diversify its economy. Nonetheless, Alaskans remain optimistic about their future. There is $23 billion in the Alaska Permanent Fund. The state possesses vast oil, fish, and expanding Pacific Rim economies.

Caribou.

Oil and Gas

The oil and gas industry continues to be the major player in Alaska's economy, responsible for creating the state's highest paid jobs in exploration and development, construction, transportation, wholesaling and business services. Additionally, it is the most important contributor to the gross state product. North Slope production has declined annually since its 1988 peak and revenue experts expect this trend to continue, with short-term production increases due to new oil-field development. Even with less production, the petroleum industry will continue to be a driver of the economy because of the many undeveloped oil and gas fields and the continued high world demand for energy. New technology will also extend the economic life of oil-field development beyond current estimated projections.

The State of Alaska is "open and ready" for gas development. Alaska North Slope's proven natural gas reserves are estimated at 35 trillion cubic feet and the probability of additional gas reserves is high. Liquid natural gas (LNG) is an environmentally friendly source of energy. Worldwide demand for LNG is growing at a rate that far surpasses the demand growth for other fossil fuels. State officials are ready to make the necessary policy decisions needed to help bring Alaska North Slope LNG to market.

Alpine Pipeline.

Chapter Four: A Vibrant Economy

The seafood industry is the backbone of economies of many coastal communities in Alaska. The state has been a strict caretaker of its fisheries. Therefore, many of the problems that have arisen elsewhere in the world, such as over-fishing, have not occurred in Alaskan waters.

Chapter Four: A Vibrant Economy

Seafood

The seafood industry, which includes fishing and processing, has been a cornerstone of Alaska's economy since the first salmon cannery was built near Ketchikan in the late 1800s. Although salmon harvesting and processing techniques were primitive compared to today's rapidly advancing technology, fishing weirs, barricades, small netting with throw nets, and beach seines were effective enough to earn Alaska its reputation for being the salmon capital of the world.

Today, Alaska's seafood industry is among the best-managed and healthiest in the world. While many of the world's fisheries suffer the irreversible consequences of over harvesting and pollution, Alaska is regarded as a world leader for implementing innovative harvest management policies and practices that help ensure it will continue to thrive for future generations. Alaska's commercial salmon industry was the first fishery in North America to receive the coveted Marine Stewardship Council's Sustainability Seal in addition to being certified an "organic" food by the USDA

If Alaska were a nation, it would rank as one of the top ten suppliers of seafood worldwide. Approximately five billion pounds of seafood are harvested off Alaska coasts each year, with a first wholesale value of $2 billion, and is equal to about 60 percent of all U.S. production. Fishing is especially important in coastal communities throughout Alaska, with most permit holders living in Dutch Harbor, Kodiak, Homer, Sitka, Petersburg, Ketchikan, and Anchorage—known as Alaska's largest fishing village with 746 commercial permit holders and 13 processors. Each year, the seafood industry pumps $3 billion into the state's economy, making it the largest private-sector employer in Alaska. One-fifth of Alaska employment is attributed to the seafood industry, and revenues generated by Alaska's seafood industry nearly equal the entire state operating budget.

Despite this flourishing resource, Alaska's seafood industry faces troubled times; the combined result of overcapitalization, changing markets, and increasing competition. As it adapts to lower prices and new management regimes, profound and lasting changes in the harvesting, processing, and marketing sectors will continue to restructure the Alaska seafood industry into the twenty-first century.

As this relic is testament to, Alaska has had quite a history with the mining industry. The minerals industry is, in fact, emerging once again as a dominant economic force in the state. Alaska mines are some of the newest, largest, and most productive in the world.

FOREST PRODUCTS

The forest products industry has been an important contributor to the economy of Alaska for over half a century. Alaska's forests are divided into two types: coastal and interior. Coastal forests are dominated by Western Hemlock, Sitka Spruce, and other softwood trees. Interior Alaska is vast with extensive stands dominated by White Spruce, Birch, and Poplars. High-quality Sitka Spruce and Western Hemlock have been exported as logs, lumber, and timbers into the Pacific Rim for the past 40 years. The lower-quality portion of the timber was used to produce dissolving pulp that was sold around the world for producing rayon, pharmaceuticals, and fine-quality paper products.

Approximately 150 commercial sawmills and secondary manufacturers operate across the state. Recent years have been tough ones for the industry, but significant changes are moving it toward value-added processing and long-term sustainability. Southeast Alaska's two pulp mills, in Sitka and Ketchikan, closed in 1993 and 1997, respectively. The Tongass Land Use Management Plan (TLMP), issued in 1997, significantly reduced allowable harvest levels, and most Asian markets are experiencing dramatic downturns in demand and price. However, many Alaskans are responding with a new entrepreneurial focus on value-added processing, and the fundamentals that created a vibrant industry, a world-class resource and a skilled, productive workforce, remain in place.

Ownership of commercial timberlands is more concentrated in Alaska than the rest of the United States. The federal government owns approximately 65 percent of all the lands in Alaska. The State of Alaska owns another 24.5 percent. Regional and village Native corporations collectively own approximately 10 percent of Alaska, and the remaining lands, less than one percent, are controlled by various private interests. Given these percentages, it is no surprise that private non-industrial timberland owners play a small role in supplying the Alaska industry.

MINING

Mining is emerging as a dominant economic force in Alaska. Strategically situated on the Pacific Rim, the state of Alaska offers prospective land, sanctity of title, a state-sponsored geological and geophysical mapping effort, a reasonable permitting process, a talented workforce, exploration incentives, and inventive infrastructure equity-sharing programs. The combination of a large base of prospects, increasing demand, and technological advances will mean increased mining in Alaska, with the mining industry expected to add between 350 and 600 jobs to the state's economy over the next several years.

Zinc, lead, copper, and silver currently lead Alaska mineral production, with lode gold, placer, and coal being the growth areas for the Alaska minerals industry. Alaska mines are some of the newest, largest, and most productive in the world, including Red Dog Mine in northwest Alaska, Greens Creek Mine in Juneau, and Fort Knox, north of Fairbanks. Although there have been recent declines in metal prices and subsequent decreases in exploration and development investment, production at both Red Dog and Fort Knox is expected to increase, and news of larger-than-expected reserves at the Donlin Creek prospect offer promising news for mining there.

Trans-Alaskan Pipeline, Section 3: Constructed by H. C. Price Co.

Chapter Four: A Vibrant Economy

In Juneau, the tram rises high above a pair of cruise ships.

Tourism

A visit to Alaska is a fulfillment of a life-long dream for many travelers. The state's scenic beauty, wilderness setting, and wildlife continue to attract and enthrall visitors. In addition to being one of the state's top two employers, non-resident tourism in Alaska is a growing economic sector. Over 1.5 million people visit Alaska each year, with 83 percent of these visitors coming during the summer season, from May to September. This is more than double the number of summer visitors in 1991, which was 727,000, and represents an average annual growth rate of about 6.5 percent.

While visitor numbers to Alaska continue to increase, the rate of growth for summer visitation has fluctuated and slowed substantially since the mid-1990s. The overall growth in the summer visitor market slowed to less than half a percent per year between 1999 and 2001, increasing to 5 percent growth between 2001 and 2002. Virtually all visitor growth during this period is attributable to the cruise sector with non-cruise sectors experiencing declining growth rates.

Transportation

Transportation plays a much larger role in Alaska's economy than in much of the rest of the nation. Transportation jobs in Alaska depend on the overall size of the economy, as well as the level of resource development and production, tourism, and construction. The state's economy and population are both expected to grow, further strengthening the transportation industry. Currently, about 10 percent of the state is employed in transportation jobs, including those attributed to the oil and tourism industries. Past growth in the transportation industry has closely followed overall growth in the economy, and those trends are expected to continue.

Most activity in the transportation industry results from activity in Anchorage, the state's largest transportation hub. Anchorage is home to the Ted Stevens Anchorage International Airport, one of the busiest cargo airports in North America with over 560 transcontinental cargo flights weekly. It is quite common to see as many as 30 wide-body jets

A visitor favorite is the "flight-seeing" tour, where small planes and helicopters transport passengers to locations that are not easily accessible, and not something to miss! Anglers also take advantage of these services to take them to the countless pristine locations for truly satisfying sport fishing.

on the tarmac at any given time. The airport's economic effects are felt as far away as North Pole, Alaska where over 100 rail cars of jet fuel are refined each day and transported by the Alaska Railroad to Anchorage.

The Port of Anchorage plays a key role in statewide transportation. It is where over 90 percent of durable consumer goods shipped to Alaska arrive. Later these items are distributed throughout the state by truck, rail, and plane. Many of the perishable goods sent to Anchorage are transported via the Alaska-Canadian Highway. The Port is served by two major carriers, TOTE and Horizon Lines, and is open and ice-free year round.

The 83-year-old Alaska Railroad is the last full-service railroad in the United States, moving 1.4 billion pounds of cargo, including importing supplies to the Interior region, exporting coal from the Interior to the Pacific Rim, and hauling pipe and equipment for the oil and gas industry. The railroad also carries 480,000 passengers on approximately 530 miles of track throughout the rail belt. In the summer months, glass-domed train cars are a top visitor attraction, transporting tens of thousands of passengers to Fairbanks, Denali National Park, Seward, Whittier, with some of the state's most spectacular scenery.

At Ted Stevens Anchorage International Airport, a new passenger depot, completed in 2003, makes transfers between planes and trains more efficient. An underground pedestrian tunnel connecting the main concourse and railroad tracks is elevated on a concrete trestle over the airport parking lots and access roads below, meeting with the terminal at a covered elevated pedestrian platform.

Longshore crews tie up one of the two container vessels that call at the Port two days a week during the winter and three days a week during the summer.

Chapter Four: A Vibrant Economy

The Alaska Railroad is the last full-service railroad in the United States.

International express cargo carriers FedEx and UPS sort packages and clear U. S. Customs for the burgeoning Pacific Rim trade.

Chapter Four: A Vibrant Economy

Alpine Forget-me-not.

Certainly this moose has no idea that he is the official state land animal in Alaska.

Chapter Four: A Vibrant Economy

Dog mushing is Alaska's official state sport.

ALASKA FUN FACTS

Motto: *North to the Future!*
Nickname: *The Last Frontier. Also, the name "Alaska" is derived from the Aleut word "Alaxsxaq" or "Alyeska," meaning "The Great Land," which is another nickname for the state.*
Capital: *Juneau*
Population: *626,932*
Area: *586,412 square miles—that makes it the largest state by far and two-and-a -half times larger than Texas!*
Flower: *Alpine Forget-me-not*
Tree: *Sitka Spruce*
Bird: *Alaska Willow Ptarmigan*
Land Mammal: *Alaskan Moose*
Marine Mammal: *Bowhead Whale*
Fish: *King Salmon*
Insect: *Four Spot Skimmer Dragonfly*
Fossil: *Woolly Mammoth*
Gem: *Jade*
Mineral: *Gold*
Sport: *Dog Mushing*
Flag: *Eight gold stars, representing the Big Dipper and the North Star, on a blue field. Designed by seventh-grader Benny Bensen for a Territorial Flag contest in 1026.*
Song: Alaska's Flag *by Elinor Dusenbury (music) and Marie Drake (words).*

Alaska's Flag

Eight stars of gold on a field of blue,
Alaska's flag, may it mean to you;
The blue of the sea, the ev'ning sky,
The mountain lakes, the flow'rs nearby;
The gold of the early sourdough's dreams,
The precious gold of the hills and streams;
The brilliant stars in the northern sky,
The "Bear," the "Dipper," and shining high,
The great North Star with its steady light,
O'er land and sea a beacon bright,
Alaska's flag to Alaskans dear,
The simple flag of a last frontier.

Chapter Five

Celebrating Alaska

Alaskans love to celebrate! Having fun comes with the territory in the Far North. And, there are many reasons for celebration. In addition to Alaska's majestic scenery, wildlife, people, history, culture, and traditions, there is Golden Days in Fairbanks, Kodiak Crab Festival, Beaver Round up in Dillingham, Talkeetna Moose Dropping Festival, Golden North Salmon Derby in Juneau, the Great Alaska Shootout, Petersberg's Little Norway Festival, Iditarod Trail Sled Dog Race, and Fur Rendezvous, Alaska's Premier Winter Festival. We truly enjoy sharing these good times with visitors, so please come join us! It's a celebration year-round and there's something for everyone!

100 Words by
Mary Pignalberi
Anchorage Fur Rendezvous

Mooses Tooth is part of the Alaska Range and is located in Denali National Park.

Celebrating Alaska

In Alaska, especially, festivals and special events take on an even bigger meaning. Some of them even started in the community out of necessity: to establish a marketplace to sell furs from the winter trap-line harvest, to get lifesaving medicine from Anchorage to Nome, or to raise money for a local non-profit.

There are practical reasons for hosting these annual events. In many Alaska communities, entertainment venues like movie theatres, galleries, and shopping districts are non-existent. Having events to look forward to depends on the interest and volunteer efforts of locals. In communities across the state, Alaskans are always finding a reason to come together, a reason to honor their way of life, and a reason to celebrate the things that are special about where they live.

From the outlandish and downright crazy events to internationally recognized competitions and festivals, Alaska's unique history and culture is celebrated in some of the most unlikely ways and in the most out-of-the-way places. Activities related to the commercial fisheries and Native Alaskan subsistence activities keep most Alaskans occupied in summer months, so most major social events are held in the heart of winter. The following are just a few of the many events that Alaskan communities have to offer.

Iditarod Trail Sled Dog Race

Unlike any competitive event in the world, the Iditarod Trail Sled Dog Race, known as "The Last Great Race on Earth," is a 1,150-mile race that covers the most unforgiving, most beautiful terrain in Alaska. Begun in 1973, the race commemorates a successful emergency mission to get diphtheria vaccine to Nome in 1925 during an epidemic. Every year in early March, about 75 mushers and their dog teams compete against each other and Mother Nature, to be the first to make it from Anchorage to Nome. Jagged mountain ranges, frozen rivers, dense forest, desolate tundra, or miles of windswept coast wait around each bend. Added to that are temperatures far below zero, winds that can cause a complete loss of visibility, long hours of darkness, and treacherous climbs, and the result is Alaska's most celebrated special event.

Alaska marmot.

This young clown just might be having too much fun at Anchorage Fur Rendezvous.

Chapter Five: Celebrating Alaska

The numerous events that take place during Anchorage Fur Rendezvous range from the sporting, such as the snow-machine race; to the playful, such as the parade; to the traditional, such as the blanket toss.

Chapter Five: Celebrating Alaska

Anchorage Fur Rendezvous

Fur Rendezvous began in 1935 as a winter sports tournament and a citywide party to help relieve cabin fever by bringing the community together and celebrating the end of winter. Because fur trading was the second-largest industry in Alaska at the time, it became an important part of the festival, providing a golden opportunity for trappers and buyers to meet in Anchorage to ply their trade and cut out the middleman. Trapping contests were held, and prizes were awarded for the longest fox, the best fox, and the finest ermine pelts. In 1937, the festival was officially named the Anchorage Fur Rendezvous. Today, "Rondy" is a week-long celebration that attracts Alaskans statewide and includes Native arts and crafts, photography competitions, fireworks, ice bowling, a parade, a sports-car grand prix, professional sled-dog racing, a Miners and Trappers Ball, crowning of a Rondy king and queen, and many other sanctioned events. In 1956, Greater Anchorage, Inc. was formed to take over the management and operation of the festival, and the non-profit organization is governed by a board of directors and has a full-time, year-round staff. Today, Rondy is one of the largest winter festivals in North America with estimates of over 67 percent of the population participating.

Juneau Gold Rush Days

Organized by the Juneau Gold Rush Commission, Juneau's Gold Rush Days celebration is an annual event that originally began as a miner's picnic featuring traditional mining competitions like jack-leg drilling, hand mucking, and spike driving. Now, Gold Rush Days is a major logging and mining event after loggers joined in the celebration in 1993, competing in a variety of events such as logrolling, log chopping, and speed climbing.

A performance by the Anchorage Opera.

Kodiak Crab Festival

Kodiak, home to the largest fishing fleet in North America, has been hosting this four-day annual Crab Festival every May since 1958, when this Southwest Alaska Island was the king crab capital of the world. Since then, the commercial fishery has declined, but the festival has continued to flourish and today attracts 15,000 visitors from around the globe. Several booths sell fresh king crab, piping hot right out of the pot, or piled on tasty crab sandwiches. Events include a halibut derby, a grand parade and a shrimp parade, a marathon, an art show, Alutiiq games, blessing of the fleet, a fishermen's memorial service, and possibly the nation's only survival-suit race, where teams wear the special suits that boat crews wear to delay hypothermia if they fall overboard, and then swim to small boats in the harbor.

Nenana River Ice Classic

The Nenana River in the Interior of Alaska starts to freeze over in October, and the ice continues to grow throughout the winter until it is several inches thick. The Nenana River Ice Classic competition began in 1917 when railroad engineers bet a total of $800, winner takes all, guessing the exact time (month, day, hour, minute) ice on the Nenana River would break up. Each year since then, Alaska residents statewide have bought about $300,000 worth of $2 tickets to guess the timing of the river breakup. A tripod, connected to an on-shore clock with a string, is planted in two feet of river ice during river freeze-up in October or November. The following spring, the clock automatically stops when the tripod moves as the ice breaks up. The time on the clock is used as the river-ice breakup time. Generally, the Nenana river ice breaks up in late April or early May, and the purse is divided by the number of lucky ticket holders.

Polar Bear Jumpoff Festival, Seward

The Seward Polar Bear Jumpoff Festival is a weekend-long winter festival in January where costume-clad Alaskans plunge into Resurrection Bay. Jumpers find sponsors to raise money for the Alaska division of the American Cancer Society, and donated prizes are awarded to the jumper who raises the most money. In addition to the plunge, the weekend is built around a full festival schedule that includes a sled-dog race, children's games, ice bowling, a seafood feed, an arts-and-crafts fair, an oyster-slurping contest, and a parade.

Tesoro Iron Dog Snowmachine Race

The Iron Dog, which is the world's longest snowmachine race, started in 1984 when a local enthusiast challenged a small field of racers to race from Big Lake to Nome, just for the fun of it. The rules were different then. Racers could ride any kind of snowmobile they wanted; they were allowed to take any parts they might need with them on the trail; and no spectators or racers were allowed to offer assistance. Today, the Iron Dog requires all racers to compete on a machine with an engine no larger than 500 CCs. Replacement parts can be stashed at checkpoints along the route and spectators and other racers can render help. A professional purse of $100,000 is offered to the top ten finishers. Although the route has changed several times over the years, today the course goes from Wasilla to Nome, then to Fairbanks, covering 2,000 miles of grueling terrain in harsh winter conditions and at speeds exceeding 100 miles an hour.

The Alaska Center for the Performing Arts provides a forum for a wide range of performances, including those of the Anchorage Opera.

Chapter Five: Celebrating Alaska

A reveler joins in the fun at the Petersburg Little Norway Festival.

World Ice Art Championships

Working with chainsaws, chisels, and drills, sculptors turn huge 8-by-5-foot 7,000-pound chunks of Fairbanks ice into everything from abstract art to wild animals every winter during the city's annual ice-sculpting competition. The World Ice Art Championships, an international sculpting competition and exhibition, produces dozens of frozen sculptures around the city and at the Ice Park, transforming the center of town into a world of ice figurines that are frozen in time.

Fairbanks has become a hotbed for ice carvers because of the superior quality and plentiful supply of ice. The championships take place in early March and host about 200 international sculptors from around the world, coming from as far away as Japan, Germany, Poland, Sweden, Mexico, and Korea. There are three different ice-sculpting events. In the single-block classic event, each two-person sculpting team is given two days and a 7,200-pound block of ice, measuring 5 by 8 by 3 feet. In the large-sculpture classic event, teams of four or five sculptors are given 12 blocks of ice measuring 4 by 4 by 3 feet. Each block weighs 3,000 pounds. In the open event, beginning sculptors or interested locals wishing to try their hand at sculpting are given a 2,700 pound block of ice measuring 3 by 5 by 3 feet, and three days in which to complete their sculpture. The artists craft the blocks into abstract, representational, or realistic sculptures, and the top three winners in each category receive gold, silver, and bronze medals.

Bering Sea Ice Golf Classic

The rules of golf are a little different in Nome, Alaska. Snow and ice divots do not have to be replaced. No swimming allowed. Beware of crab-fishing holes and blowing snow. If the ball hits a polar bear, the player loses three strokes. If he gets the ball back from the bear, five strokes will be subtracted from his final score.

The Bering Sea Ice Golf Classic is held in March as part of Nome's month of activities to celebrate the town's fame as the finish line of the Iditarod Trail Sled Dog Race. The six holes are laid out on the Bering Sea ice just south of town. Each one is made from a flag and a coffee-can set into spray-painted greens. The view from this course is truly one of a kind. A repository of Nome's Christmas trees provides the habitat for plywood cutouts of Alaska's big game, including moose, bears, and musk oxen.

An entry in the snow-carving contest at Fur Rondy.

The course is par 41, although it has been won in 23 strokes. A hole-in-one requires a shot of at least 120 yards, though the Lions Club warns players that on the ice, even a soft tap on one of the fluorescent balls can send it 200 yards. They also recommend that players leave the good clubs at home and use clubs supplied by the course instead.

THE NATIVE YOUTH OLYMPICS

Attended by over 100 teams from high schools statewide, the Native Youth Olympics are competitive events and cultural exchanges held in Anchorage every April. Native Alaskan youth from across the state share the eight event games with others, and at the same time teach spectators about the traditional contests of their forefathers. The origins of each event are based on traditional Native Alaskan activities. For example, the Stick Pull originated from hunting activities. A successful hunter has to be able to pull a seal out of the water, requiring balance and strength. The Scissor Broad Jump is based on the ability of hunters to jump from one ice floe to another while maintaining balance on melting, shifting and breaking ice. The One-Hand Reach is a game to test a person's control over their body. If a hunter was to become lost in open water, they must know the skills to control their body in order not to panic and tip the kayak. The Seal Hop, a game of sheer endurance to see how far a person can go on pure determination, originated from the hunter imitating the movement of a seal during the hunt. Organizers contribute the success of the Native Youth Olympics to the uniqueness of the event and the athletic style, and to the effect the spirit of sportsmanship has on the participants and spectators.

A little Norwegian wanders the beach at the Petersburg Little Norway Festival.

Chapter Five: Celebrating Alaska

Re-enactment of the transfer ceremony atop Castle Hill in downtown Sitka where on October 18, 1867 the United States took possession of the Russian American Company.

Courtesy of the City and Borough of Sitka; Photo by Will Swagel

A mysterious reveler enjoys Fur Rondy's Masque Ball.

Chapter Five: Celebrating Alaska

Courtesy of Anchorage Fur Rendezvous

Chapter Five: Celebrating Alaska

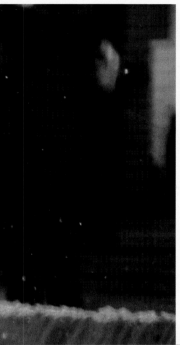

After seeing dramatic photos of both past and present challenges, it is easy to understand why the Iditarod is considered The Last Great Race.

The Last Great Race on Earth

The very air is brisk and exhilarating. The dogs are bright-eyed and are leaping with excitement. The mushers are determined and ready to go. It is the start of the Last Great Race on Earth—the Iditarod Trail Sled Dog Race.

The race commemorates a stunning achievement when, in the winter of 1925, Nome was struck with an epidemic of diphtheria. Serum was desperately needed, but transportation options were limited. The serum was transported from Anchorage to Nenana by train. Then, the much-needed serum was raced to Nome by a series of dog teams that relayed the medicine 674 miles in 127.5 hours.

Today, the Iditarod celebrates the accomplishment of those determined mushers and dedicated dogs. The competition begins on the first Saturday in March with a ceremonial start in the heart of Anchorage. The next day, the teams start a course that begins in Wasilla and covers approximately 1,150 miles to Nome. Each year, the course switches between a northern route, which is used on even-numbered years and a southern route, which is used on odd-numbered years. Mushers must check in at each of 26 checkpoints and are required to make at least three mandatory stops: one for 24 hours and two for 8 hours. The bond between the musher and his or her dog team is formidable, and the musher's most important duty is to care for the animals. Strict rules regarding the treatment and well-being of the dog teams are imposed, enforced, and willingly complied with.

About nine to eleven days after the start of the competition, the first musher arrives in the city of Nome. For the next ten days or so, teams continue to arrive, day and night. Every musher is greeted by a fire siren and a crowd lining the "chute," as the team runs down Front Street to the burled arch.

All manners of events and celebrations accompany the Iditarod in Anchorage, Nome, Wasilla, and all the communities along the route. It is a thrilling and boisterous time to be in Alaska, celebrating with Alaskans.

Chapter Six

AMERICA'S LAST FRONTIER

An untold number of people have attempted to describe Alaska. None do it justice. Incredibly diverse and geographically complex, my state defies a single superlative.

Though we cannot adequately describe its wonders and riches, we Alaskans are dedicated stewards of the Last Frontier who recognize the absolute necessity for wise decision making. The conservation and development choices we make today will affect our state and our people for decades to come.

I am proud to be an Alaskan. We take seriously our responsibility to preserve opportunities for future generations as we develop our state's rich resources. We will choose wisely.

100 Words by
Drue Pearce
U.S. Department of the Interior

Shishaldin Volcano, elevation 9,373 feet, is currently one of the most active volcanoes in the Aleutian arc. Shishaldin is located on Unimak Island, in the eastern Aleutians. False Pass, the nearest village, is 20 miles away.

America's Last Frontier

For hundreds of years, artists have tried to describe Alaska's natural beauty. Each of them has had the same limited success. There are simply no words to accurately capture the beauty of Alaska. No amount of paint can accurately portray the pallet from which Mother Nature has designed the landscape.

Alaska is several lands in one. The distances are so great and the diversity of the terrain and climate so extreme that, in an 1869 speech, Secretary of State Seward predicted the region might have to be admitted to the Union as several separate states.

Alaska—the Great Land—everything is larger than life. The entire state covers 367 million acres, by far the largest state in the union and one-fifth the size of the Lower 48 states. The record-breaking cabbage grown in Alaska tipped the scales at 105.6 pounds, and a blue ribbon rutabaga weighed in at 75 pounds, nurtured in part by the almost 20 hours of daylight that shines long after the day is done in the rest of the nation.

Of Water and Ice

Of course, there is water, in all its glorious forms, carving and texturing the topography. There are 3,000 rivers, with the Yukon being the longest. It originates in Canada and flows 900 miles before snaking along the last 1,400 miles in Alaska and finally emptying into the Bering Sea. Alaska has more than 3 million lakes that are more than 20 acres in size.

From the cut-out coves of Southeast Alaska, to the frigid shores of the Arctic Ocean, 6,640 miles of coastline hugs the shores that are home to many Alaskans. There are 1,800 named islands—Kodiak Island being the second largest in the United States after Hawaii—and several thousand more unnamed islands.

Much of Alaska's water is locked in ice. Massive ice fields and glaciers cover approximately five percent of the state or 29,000 square miles. Glaciers are formed, where over a number of years, more snow falls than melts. Glacier ice often appears blue to the eye because it absorbs all the colors of the spectrum except blue, which is reflected back. There are 100,000 glaciers ranging from tiny cirque glaciers to huge valleys of ice. About 75 percent of all fresh water in Alaska is stored as glacial ice, many times greater than the volume of water stored in all the state's lakes, ponds, and rivers.

Killer whales.

Columbia Glacier.

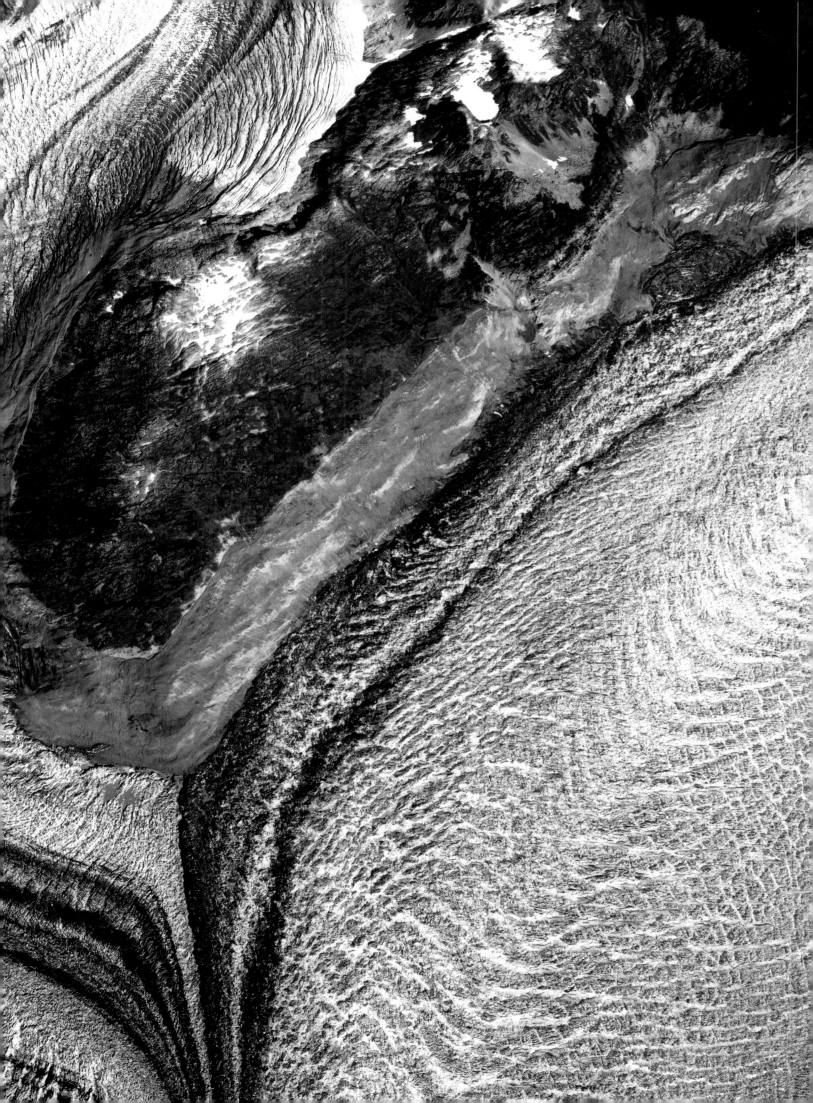

Chapter Six: America's Last Frontier

NOAA's Little Port Walter Facility, located near the southeastern tip of Baranof Island, is the primary field research facility of the Auke Bay Laboratory. Accessible only by boat or seaplane, Little Port Walter is the oldest year-round biological research station in Alaska and has hosted numerous fisheries research projects since 1934.

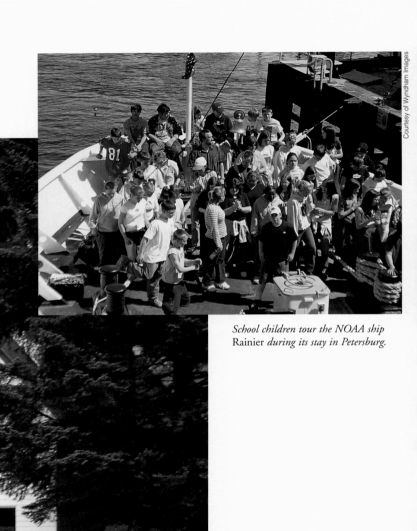

School children tour the NOAA ship Rainier *during its stay in Petersburg.*

MARINE WILDLIFE

The marine wildlife that relies on many of these critical habitats for safe harbor attracts visitors from around the world and amazes even the most seasoned Alaskans. As the ice melts in the northern waters of Alaska each spring, gray whales are frequently sighted in large numbers from Alaska shores in Southeast, Kodiak, Seward, and the Bering Sea, as they migrate from California to Alaska. Gray whales have the distinction of being the most ancient of the living "mustached" or baleen whales. Migrating between Baja California and the Bering and Chukchi Seas, they travel a round-trip distance of 10,000 miles, one of the longest migrations of any marine mammal. Gray whales are recognized by their mottled gray color, scars, abrasions, and clusters of barnacles on their heads and backs. In just the five months spent in Alaskan waters, an adult gray will eat almost 400,000 pounds of food, feeding on the bottom by sucking tube-dwelling amphipods and other invertebrates out of the sandy sentiment and leaving large oval feeding imprints behind.

Humpback whales provide some of the most exciting photographic moments of all the marine mammals. Professional and amateur photographers from around the world come to Alaska to capture peak moments such as these on film: the mighty launch rockets 40 tons of twisting whale skyward and pectoral fins fling sparkling sheets of water as a 45-foot-long body lifts almost entirely out of the water in a spectacular breaching display. This is a frequent occurrence in the summer months along the Inside Passage of Southeast Alaska, where one of the world's greatest populations of humpback whales resides in these nutrient-rich waters. Besides breaching, they often slap their mighty tails repeatedly or clap their 15-foot pectoral fins in the same fashion.

Although some populations of the beluga, or white whale, are strongly migratory, there are a few isolated populations that do not migrate in the spring, including those in Cook Inlet, Alaska. They can be spotted in Turnagain Arm from downtown Anchorage, as well as in Bristol Bay, Norton Sound, Kotzebue Sound, and Kasegaluk Lagoon. Known as "sea canaries" for being very vocal, they produce a variety of grunts, clicks, chirps, and whistles that are used for navigating, finding prey, and communicating. The beluga is also the only whale that can bend its neck. This helps them to maneuver easily and catch prey using their teeth, not for chewing, but for grabbing and tearing their prey that is then swallowed whole.

Former Mayor of Sitka Valorie Nelson joins a NOAA crewmember on an icefield survey in Southeast Alaska.

Black bear.

Orcas, also known as killer whales, are the largest member of the dolphin family and are called "killer whales" because some groups attack and consume whales and other large prey such as seals and sea lions. They can be found statewide, in the waters from Southeast Alaska through the Aleutian Islands and northward into the Chukchi and Beaufort seas.

The most recognizable feature of the orca is the high dorsal fin that serves the whale as a keel does a boat and can grow to six feet in height on males. They are regarded for their intelligence and possess all typical mammalian senses except smell.

The Great Ones

On land, too, Alaska's natural world is humbling. The state has 39 mountain ranges, containing 17 of the 20 highest peaks in the United States. Mt. McKinley, within the Denali National Park and Preserve, is in the Alaska Range and is the tallest peak in North America at 20,320 feet. The mountain was named for William McKinley of Ohio, at the time a Republican candidate for president. Denali, the mountain's indigenous name, is an Athabascan word meaning "the high one" or "the great one."

Together, the mountain and the park are one of the top visitor attractions in Alaska. Although August is the best month to see McKinley close up, the mountain is rarely visible the entire day, and it is not unusual to be able to see the mountain clearly in Anchorage, almost 250 miles away, when it is completely blocked by its own weather conditions just miles from its base.

The Chugach, Wrangell, and St. Elias mountain ranges converge in what is often referred to as the "mountain kingdom of North America." The largest unit of the National Park System and a day's drive east of Anchorage, the Wrangell-St. Elias National Park and Preserve includes the greatest collection of peaks above 16,000 feet. Mount St. Elias, at 18,008 feet, is the second-highest peak in the United States. Adjacent to Canada's Kluane National Park, the site is characterized by remote mountains, valleys, wild rivers, and a variety of wildlife. The continent's largest assemblage of glaciers is also found here.

Wildlife Viewing

Truly one of the most magnificent places to watch wildlife in all of North America, Denali National Park is home to a number of grizzlies, gray wolves, moose, caribou, golden eagles, Dall sheep, and more. It is also is a land of extremes, with long days in June that have almost no darkness and a short fall season with color that only lasts about two weeks and is usually in full swing by the last week of August. It is not unusual for snow to have closed the park as early as mid-September.

Wrangell-St. Elias has been targeted by the state as the site that is expected to be the next most popular destination for visitors. While there is also a vast amount of wildlife in Wrangell-St. Elias, opportunities to view it are limited due to dense brush and forest along the roads. The best spots for viewing wildlife are from alpine areas above tree line. Wrangell-St. Elias contains one of the largest concentrations of Dall sheep in North America, which can be easily spotted dancing

Wildlife viewing is one of the joys of travelling through Alaska. Sometimes, the wildlife view back.

Chapter Six: America's Last Frontier

No artist's brush can convey the light and colors of Alaska.

along rocky ridges and mountainsides. Moose are often seen near willow bogs and lakes. In the fall, bears and other animals may be sighted near salmon spawning streams. Other species of large mammals include mountain goats, caribou, moose, brown/grizzly bear, black bear, and two herds of transplanted bison. The coastal areas of the park are habitat for marine mammals, including sea lions, harbor seals, sea otters, porpoises, and whales. Small mammals include lynx, wolverine, beaver, marten, porcupine, fox, wolves, marmots, river otters, and other small furbearers and rodents.

Aurora Borealis

No description can relate the splendor or the magnificence of the natural phenomenon known as the northern lights or *aurora borealis*. It is in the northern regions of the state where the northern lights also give their most spectacular displays. A 17th-century scientist named Pierre Gassend, applied the name Aurora to the northern lights, named after the goddess of dawn in Roman mythology. Every northern culture has legends about the lights and often associates them with life after death. They have been described in ancient times by the Eskimos, American Indians, world explorers, and are even mentioned in the Old Testament. The only Eskimo group that considered the aurora borealis an evil thing were the Point Barrow Eskimos. They believed this so deeply that they used to carry knives to keep it away. The Tlingits and Eyak Indians of Southeast Alaska considered them a sure sign of impending battle and believed that someone would be killed when they saw the cosmic light display.

The Geophysical Institute of the University of Alaska is a major station for the study of the aurora borealis with specialized cameras and improved spectroscopes. It was found that the displays were caused by magnetic disturbances from the sun, which produced light when colliding with atoms in the upper atmosphere. Scientists do not deny that the aurora may cause weather changes, due to the expansion of the upper atmosphere affecting the lower atmosphere where the weather originates. The aurora borealis encircles the entire Polar Regions. Observers on Earth only see a small part of the aurora's display since the lowest sections of the aurora are 40 miles up. Astronauts looking down on the polar region from space have a better overall view to observe the aurora borealis as it extends about 600 miles above the earth!

The aurora borealis, also known as the northern lights, is a major tourist attraction.

THE GREAT ONE

To many, the majestic Mt. McKinley should more fittingly be called by its Native name, Denali—the Great One. For great it is. While Denali is the highest mountain in North America, it is arguably the single most impressive mountain in the world. Its elevation is a soaring 20,320 feet. All other higher peaks are in the greater Himalayas or in the Andes, all part of huge mountain ranges. But Denali stands alone, rising 16,000 feet above the snowline. While it is indeed part of the Alaska Range, Denali so dominates its area that what would otherwise be major ice peaks appear to be mere foothills in comparison.

Denali is unique in another way. Outside of Antarctica, it is perhaps the coldest mountain in the world, with the summit area below zero degrees Fahrenheit nearly year round and vicious winds whipping the peak incessantly. The combination of height, latitude, and weather are the contributing culprits to the frigid conditions.

The Great One is the crown jewel of Denali National Park and Preserve. With more than 6 million acres, the Park features spectacular mountains and glaciers, and encompasses a complete sub-arctic ecosystem including large mammals such as grizzly bears, wolves, Dall sheep, and moose. Visitors can enjoy a wide range of activities from wildlife viewing, to backpacking, to mountaineering. The Park also continues to be home to a laboratory for natural-science research.

Denali—imposing, mysterious, and ever-challenging—is the Grand Dame of the Alaskan Range. On a clear day, its snow-capped peak glistens brightly and shows its face clearly in Reflection Lake. It seems more than a mountain. It is the Great One.

Tundra flora.

Chapter Six: America's Last Frontier

Mt. McKinley.

Courtesy of AeroMap U.S.

Aialik Coast.

Chapter Seven

The Great Outdoors and the Sporting Life

No words of north-country literature adequately portray the great state of Alaska and its ageless beauty. No one has yet been able to record all the attributes of this largest of states. We have Olympic-caliber cross-country ski trails, a top-ten ski resort that has fostered several Olympians, year-round fishing, snowmachining, hiking, climbing, windsurfing, kayaking, rafting, boating, as well as traditional and native sports.

When in Alaska, you truly feel the boundlessness of our great outdoors. It's a place to "Dream Big and Dare to Fail." The spirit of adventure lives within all of our citizens, and visitors leave feeling renewed.

100 Words by
Norman Vaughan
Explorer

The Great Outdoors and the Sporting Life

In 1883, one of Alaska's first visitors wrote, "a round trip ticket means more unalloyed enjoyment than can be crowded into a similar two-week trip in this country or any other."

Today, Alaska is one of the most popular destinations in the world, and every year more than 1.5 million visitors travel from around the globe to be a part of the Great Land. By air, water, road, and rail, they explore the natural and cultural attractions statewide. One out of three visitors is making a return trip to the state. For many it is the trip of a lifetime; the first and last time they will ever set foot on Alaskan shores.

Until the early 1970s, Alaska was considered a barren, remote frontier, and traveling to the state was paramount to going on an exotic African safari. Few visitors dared brave the journey outside of the brief four-month season between May and September. They traveled in large, commercially organized groups, too uncertain of what was in store for the independent traveler. However, during the past decade, visitors have started to learn that the fall and winter seasons are the savvy traveler's "best kept secrets," and they are starting to arrive—in groups and independently—for skiing, dog sledding, northern lights viewing, fall fishing and hiking, and the many winter festivals and special events.

If they have time, travelers come by car or motor home via the Alaska-Canada "ALCAN" Highway, built during World War II for the war effort. Most come by air or sea. The state-owned ferry system, Alaska's Marine Highway System, has linked Alaska to British Columbia and the state of Washington since the early 1960s. Each year, thousands of ferry travelers experience the stunning sea and landscapes of the Inside Passage via the Alaska Marine Highway. Since the 1970s, the cruise-ship industry has met that same growing tourist demand, and today Alaska is one of the top three cruise destinations in the world. By the beginning of the 21st century, the cruise-ship industry had expanded its service to many ports of call in Alaska. Currently the most highly visited port of call for cruise ships is the City and Borough of Juneau.

Dall sheep.

Ice climbing.

Chapter Seven: The Great Outdoors and the Sporting Life

Great horned owl.

Ptarmigan Lake rainbow.

Chapter Seven: The Great Outdoors and the Sporting Life

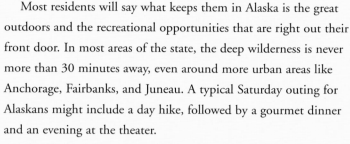

A little beachcomber near Petersburg.

Most residents will say what keeps them in Alaska is the great outdoors and the recreational opportunities that are right out their front door. In most areas of the state, the deep wilderness is never more than 30 minutes away, even around more urban areas like Anchorage, Fairbanks, and Juneau. A typical Saturday outing for Alaskans might include a day hike, followed by a gourmet dinner and an evening at the theater.

The Alaska State Park system is comprised of seven state parks: Chugach, Denali, Chilkat, Kachemak Bay, Point Bridget, Shuyak Island, and Wood-Tikchik—the latter being one of Alaska's most remote state parks. With 1.5 million acres of land, it is the largest contiguous state park in the United States. The 49,000-acre Alaska Chilkat Bald Eagle Preserve has the world's largest concentration of bald eagles each fall. In addition to camping and picnicking, these parks are ideal for hiking trails, boating, fishing, hunting, canoeing, kayaking, and rafting, along with cultural, natural and historical resources. Winter activities include snowshoeing, snow machining, cross-country and backcountry skiing, winter camping, ice fishing, and wildlife viewing.

Alaska has two national forests, the Tongass and the Chugach. The Tongass, which covers 17,000,000 acres in Southeast Alaska, is the largest, standing old-growth forest in North America. The Chugach is one of the nation's wildest national forests, and has been continually inhabited by Alaska Native peoples for more than 10,000 years. Today, the Chugach is still home to people, brown and black bears, eagles, salmon, hundreds of bird species, moose, wolves, mountain goats, deer, whales, sea otters, fox, and a countless number of other animal species. It has over 400 campsites that can accommodate 2,000 people, more than 40 public-use cabins available year-round, and well over two million acres of backcountry open for camping where no permit is required. More than 200 miles of trails, as well as endless off-trail possibilities, exist throughout the Chugach.

Alyeska Ski Resort, just 45 minutes south of Anchorage in the quaint ski village of Girdwood, is the state's largest full-service ski resort. Alyeska attracts skiers and snowboarders from around the world to recreate in the surrounding Chugach National Forest during the winter months. Home to Olympic gold medallist Tommy Moe, it is said by skiers who know the sport "if you can ski Alyeska, you can ski anywhere in the world" because of its steep terrain. Along with the Eaglecrest Ski Area in Juneau, it is one of the few resorts in the world where skiers have a view of the ocean as they make their way down the mountain.

Walrus.

Although it is not for everyone, this section of the Chugach is also known for extreme sports such as heli-skiing and snow-cat skiing where daredevils are guaranteed thousands of vertical feet and unlimited powder runs in an afternoon. Skiers are transported to the top of the mountain via helicopter or snow cat for the envied opportunity to ski or snowboard down virgin powder snow blanketing the mountainous terrain known for its gentle alpine bowls, giant mountain faces, massive glaciers, and sheltered tree skiing.

Valdez, where the waters of Prince William Sound define the shoreline, is the sight of one of the most extreme winter events in the world. Every year, about 30 elite, competitive free-skiers come from around the globe to ski down as much as a 4,000-foot vertical drop during the World Extreme Skiing Championships. Because of changing weather, contest officials wait until the last moment to determine where the contest begins. Competitors do not know until the very last minute what course they will be skiing. At that point, the athletes are allowed one fly-by peek at the course before they are dropped off at the top of the run.

Another elite sport is mountain climbing. Each season about 1,200 mountaineers from all over the world attempt to ascend the staggering heights of the 17 highest peaks in North America, including Mt. McKinley, the grand dame at 20,320 feet. About 50 percent of those attempting to reach the summit during the 110-day climbing season succeed, and the others usually turn back or are rescued because of stormy weather on the mountain, or due to personal circumstances.

Alaska is recognized for having the best fly fishing, saltwater fishing, and freshwater fishing in the world. Fishing is such a part of life in Alaska that, in many places, anglers enjoy fishing from the side of the road or streamside fishing that is just a short walk in. Serious anglers and enthusiasts can choose fly or light-tackle fishing for salmon and trout in freshwater streams and lakes; or, in clear protected saltwater fjords, one might try fishing for all Pacific Salmon, halibut, and over 20 more species of ground fish.

Charters range from small outboard motor boats, to float-plane operations that take enthusiasts to secret fishing holes where the fish are so thick anglers like to say, "you can walk across them." Five-star fishing lodges offer expert guides and state-of-the-art gear, along with meals prepared by professional chefs, homemade baked goods, and across-the-board customer service that includes cleaning and packing the day's catch.

Although stories of Alaska often portray it as a land of extremes and hard adventure, meant only for the hardiest outdoor enthusiasts, in fact, visitors of all ages and abilities are able to participate in the great outdoors. Two activities visitors and residents never tire of are sightseeing and wildlife viewing. A slight shift in the lighting embellishes the deep blue visible in glacial fields, creating an entirely different perspective. A storm moving into the area can, within minutes, completely obstruct craggy mountain peaks that normally tower above the clouds at 10,000 feet. Retreating tides leave intricately carved designs in the resulting "mudflats"—often consisting of a deadly combination of soil and sand that has many of the same properties as quick sand.

On land or water, the best wildlife viewing is in the heart of the wilderness, though it is not unheard of to see a family of moose holding up traffic as they meander through a busy intersection. Bears are often spotted in backyards, attracted to food scents from garbage cans and in search of birdseed on the ground. In coastal communities, sea lions are not shy about swimming right up to the docks, and eagles make their homes in trees adjacent to seafood-processing facilities, waiting patiently for scraps.

Chapter Seven: The Great Outdoors and the Sporting Life

Many visitors to Alaska are looking for the best of both worlds—outdoor adventure such as sport fishing, along with all the comforts of home. For such intelligent individuals, there are many facilities such as Alaska Lakeside Lodge, where one can fly off to catch salmon by day and come back to a great dinner and a warm, comfortable room by night.

This helicopter took visitors on top of this awe-inspiring glacier.

Chapter Seven: The Great Outdoors and the Sporting Life

Bantam Regional Hockey in Fairbanks.

Snow machines and their riders fly through the air as they race to the finish at Anchorage Fur Rendezvous.

Chapter Seven: The Great Outdoors and the Sporting Life

Hiking trails statewide offer a range of difficulty, from those that are flat, scenic walks, or paved biking trails ideal for seniors and children, to those that require more skill and endurance, depending on elevation, topography, and trail condition. Many of these same trails are used for biking and cross-country skiing later in the season. Sea kayaking requires little skill or endurance for an afternoon outing and gives participants access to coves and marine outlets they would otherwise have no way of reaching.

In a state where ponds, lakes, and rivers are frozen for much of the year, it only makes sense that hockey is king. Every winter, grade-school playgrounds are flooded and groomed for children to skate on in the morning before school starts and during recess. Adult hockey leagues for men and women of all ages, combined with those for children, make the sport so popular, it is not uncommon for leagues to have games at 10 p.m., or to have practices at 6 a.m., to accommodate the demand with only limited indoor facilities.

The Alaska Aces, a minor-league professional hockey team in the West Coast Hockey League, is based in Anchorage and is the only claim to professional sports Alaska can make. Although only an organized team since 1991, Alaskans support the homegrown, hometown team whose leading skaters often got their start on Alaska's winter playgrounds. Between the Aces and the University of Alaska Anchorage Seawolves (Western Collegiate Hockey Association) hockey team, more than 215,000 seats are sold for annual hockey games held in Anchorage.

Those from rural communities say that basketball is the favored Alaska sport, but instead of the stars being professional or collegiate, they are high-school girls and boys who travel statewide to play against other schools for seasonal play, tournaments, and the coveted championship. Basketball's popularity in bush Alaska is for practical reasons—although rural towns do not have athletic clubs, each community's school has a basketball court, so the sport can be played year round. Besides the logistics of team travel, often one of the biggest challenges is maintaining enrollments high enough to recruit more than just a first-string lineup.

Mt. Marathon Race in Seward.

Colorful Characters of the Last Frontier

The Legend of Soapy Smith

Certainly one of the most colorful characters in Alaskan history—and a testament to its wild frontier past—was Jefferson Randolph "Soapy" Smith. Soapy Smith earned his nickname after running a confidence game in which he sold bars of soap out of a large tub, one of which was supposedly wrapped in a $100 bill. Oddly enough, the bar with the $100 never seemed to turn up.

Soapy first turned up in Skagway in 1897 and he soon had a saloon and casino called Jeff's Place. He had a gang of helpers he called his "lambs" that saw to it that the few who managed to win any money in his casino were met in Skagway's back alleys and relieved of their winnings.

Soapy invested his earnings into Skagway's first telegraph station. There, he dutifully took money from customers eager to send wires to loved ones at home. They also received replies that invariably involved a request for money to be sent home—which of course the telegraph office was happy to help with. Few residents ever realized that the wire that ran from the office led to nowhere.

It seemed that Soapy could do no wrong however. He had the town and its officials in his back pocket. But just four days after leading the 1898 Fourth of July parade as Grand Marshall, Soapy Smith's luck ran out. Triggered by the indignation of a foolish miner who was outwitted out of nearly $3000 worth of gold, a vigilante group called the Committee of 101, with Frank Reid as its leader, went after Soapy.

It ended just like a classic western. Reid and the Committee gave Soapy an ultimatum, "give back the gold, or else!!" Soapy, ever defiant and with a grudge against Reid, refused, saying that he won the gold fairly. A struggle ensued. Soapy's last words were reputed to be, "For God's sake man, don't shoot!" Too late for him and for Frank Reid. Soapy lay dead at the age of 38 and Reid lay mortally wounded—he died 12 days later.

Today, Jefferson Randolph "Soapy" Smith assumes larger-than-life proportions in his adopted hometown of Skagway. His outrageous and colorful life has been woven into the fabric of Alaskan history.

Soapy Smith in his saloon.

Chapter Seven: The Great Outdoors and the Sporting Life

Whiskered auklet

Chapter Seven: The Great Outdoors and the Sporting Life

Courtesy of Department of Community and Economic Development

Alaska wildflower.

Chapter Seven: The Great Outdoors and the Sporting Life

Kathleen Eloisa Rockwell, also known as "Klondike Kate" to her many fans in Dawson.

Will the Real Klondike Kate Please Stand Up?

It seemed, for a time, that any woman with the name of Katherine or Kathleen who lived in Dawson, or anywhere near the Yukon, was called Klondike Kate. There are at least 17 references to different Klondike Kates in the history books. Here are brief stories of two of the most famous women.

Born in 1869, Katherine Ryan had been a nurse in Seattle. She moved on to Vancouver, British Columbia where she caught gold-rush fever and made her way to the Klondike to stake no less than three claims. Among the many accomplishments of her very colorful life, she became the first female member of the North West Mounted Police, the first female gold inspector, a jail keeper, a restaurateur, a suffragette, and more. It is said, that "Klondike Kate" Ryan's motto was, "I wasn't built for going backwards. When I once step forward, I must go ahead." Klondike Kate Ryan died in 1932.

Born in Kansas in 1876, Kathleen Eloisa Rockwell began her career as a chorus girl. She joined a vaudeville troupe and traveled from New York City to Spokane, Washington. Her job was to sing, dance, and generally encourage men to purchase as many drinks as possible. After seeing headlines about the Klondike Goldrush, Kate headed north, "buck and wing dancing" (tap dancing) her way to Dawson City in 1900. While there, she became a particular favorite, especially among the miners, with her red hair and ability to sing and dance. She was an entertainer at the Palace Grand and later at Dawson's Orpheum Theatre. Her fame as an entertainer seemed to be only somewhat less colorful than her love life. She was embroiled in a relationship with the owner of the Orpheum, Alexander Pantages, and had a string of broken marriages.

Her life after she left the Klondike in the early 1900s was decidedly less exciting, but she always managed to promote herself. Until her death in 1957, she capitalized on her life and called herself such names as "Queen of the Yukon," "Belle of Dawson," and "Klondike Queen."

North Slope Drilling Platform.

Chapter Eight

North to the Future

It is hard for many to fathom how large Alaska really is. I was fortunate to be born in Sitka and always enjoy watching the astonishment of visitors when they discover the beauty and then the size of Alaska. And yet, despite this great size comes a strong community feeling among Alaskans. We tend to look after each other. This is even more the case today with the advent of the fast telecommunications we enjoy that has brought all Alaskans even closer. Alaska is a Great Land, with Great People with Great Splendor and truly is "North to the Future."

100 Words by
Fred Reeder
Mayor-City and Borough of Sitka

North to the Future

Alaska is an exciting place to live and work—a land with a rich past and a future waiting to be discovered. Alaskans understand their involvement in both the economy and social fabric of the state makes a positive difference in the quality of life in the north.

Alaska's economy is based largely on the gifts of nature. Oil, gas, seafood, scenic beauty, minerals, and timber make up the foundation upon which the economy is built. Of all the natural resources, the oil and gas sector dominates the economic base, accounting for 49 percent of the business that creates new wealth.

Oil and Gas Industry

The oil and gas industry includes the exploration, development, and production of oil and gas products. The industry also includes oil and gas field services and pipeline transportation. In Alaska, the oil and gas industry is a major employer as well as the economic driver for the state. According to a 2001 study by the McDowell Group, the industry generates approximately 33,500 jobs and a $1.4 billion payroll annually in the state. It also generated almost $2 billion in revenues for the State of Alaska during FY 2001, with an additional $344 million in royalties going into the Alaska Permanent Fund.

The most significant trend in the state's economy is the declining production from Prudhoe Bay, which, because of its huge size, sends ripples throughout the state's economy. However, a number of other North Slope oil units have come online to supplement the Prudhoe Bay production, and the total North Slope oil production is expected to level out at about one million barrels per day through 2010. There are also enormous amounts of natural gas reserves in the North Slope reserves. Construction of a natural-gas pipeline from the North Slope is under active consideration. The gas line, estimated to cost up to $20 billion, depending on routing, enjoys strong political support. However, both prices and markets must be secure to prompt investment.

More than 70 percent of the world's population of northern fur seals gather to breed on the Pribilof Islands, an area of focus for The Nature Conservancy.

Seawater Injection Plant.

Chapter Eight: North to the Future

AIC supported the offloading of the massive production modules that were transported by barge during the summer to Northstar Production Island. Each module weighed in excess of 3,000 tons.

In sub-zero temperatures, sea-ice excavation becomes a steamy activity. The ice is being removed by a backhoe to construct an offshore gravel island.

Chapter Eight: North to the Future

Bridge construction during winter requires a significant thawing effort for the pile to be driven to the correct depth. As shown here, the pile-driving work was done at minus 40 degrees. AIC has the knowledge and equipment to install all types of pile in a wide range of conditions.

The state's second active oil and gas-producing region is in Cook Inlet. Several smaller, independent operators have succeeded to leases formerly held by major oil and gas companies. Using new exploratory and drilling techniques, these operators have re-invigorated interest in the region.

The State of Alaska is planning to explore for the availability of coalbed methane gas. The program will provide information as to the costs and feasibility of future development. Another program the state has embarked upon is a shallow gas lease program, the intent of which is to locate local sources of gas that can be delivered to rural and remote communities at less cost than alternatives.

Seafood Industry

The seafood industry is very important to both the economy and the social fabric of Alaska. Alaska's fishery management systems have evolved in an effort to maintain the rich ecosystem. The Alaska fishing industry is under constant pressure to keep up with technical advances in fishing and seafood processing and to maintain and improve its global market position. While total commercial harvests have remained fairly constant in recent years, fisheries and related activities are nonetheless areas of major growth potential for the Alaska economy. This is because many important components of the industry that could potentially be in Alaska are still located outside the state.

Nearly 49 percent of U.S. commercial seafood harvest by weight came from Alaska in 2000, equating to 27 percent of the total value. For most of Alaska's coastal communities, fishing is the backbone of the economy. During 2000, 45,550 people were engaged in commercial fishing and seafood processing in Alaska. This translated into 27,877 full-time-equivalent jobs, of which Alaska residents held only 36 percent.

The Alaska seafood industry must constantly innovate to adapt to changing marine and market environments. The ability of the industry to alter its operations to balance these dynamic forces will dictate its long-term success, and to a large extent, the health of Alaska's coastal communities. The state has identified several areas of opportunity for the seafood industry.

Fishing boats in Petersburg.

Wild vs. Aquaculture. Over the next ten years, aquaculture will be a pervasive force on the Alaska fishing industry. In order to establish the superiority of its wild seafood brand over farm-based competitors, Alaska's seafood suppliers must embrace consumer-based market strategies aimed at differentiating wild, natural products from farmed. To remain competitive, the Alaska seafood industry must continue to lower costs and increase efficiencies while increasing product value and diversifying.

Shellfish Aquaculture. Although Alaska does not permit finfish farming, it is legal to raise shellfish in the state. A growing number of aquatic farms in Alaska raise shellfish including oysters, mussels, and clams, and the industry is gearing up to introduce additional species.

Resident vs. Non-Resident Harvest. A large majority of fisheries earnings go to people residing outside of the state. Increasing participation by Alaskans in the harvesting and processing of Alaska's bountiful seafood resources is a major challenge facing the state.

Sustainability. The overall health of fish stocks in Alaska is excellent. However, unless more is done to get the word out to consumers about Alaska's enviable harvesting and management methods, the Alaska seafood industry faces being tainted with "guilt by association" due to the poor fishing practices of some other fishing nations. Compounding the problem at the national level is a trend to reduce federal resources for commercial fisheries development as attention turns to solving overfishing issues in other parts of the country and promoting domestic aquaculture operations.

Bycatch Reduction. "Bycatch," or the non-targeted fish species incidentally caught in a particular fishery, has been significantly reduced with major improvements. Innovations in management practices and development of improved gear, coupled with market incentives and regulatory requirements, will drive continued gains in this area.

Other areas of opportunity include fishery management practices that aid in the protection of endangered species, greater utilization of harvested fish and shellfish and reduction of waste, market diversification, expansion of the individual fishing quota concept and the discouragement of the "race for the fish," improved product development, and increased lobbying efforts to encourage the federal government to recognize seafood in the same way as livestock and other protein sources for the purposes of several federal food programs.

Tourism Industry

The visitor industry has a growing role in the Alaska economy. In addition to being one of the state's top employers, non-resident tourism in Alaska represents a growing economic sector.

During the 2000-2001 visitor season, over 1.4 million people visited Alaska—83 percent of them during the summer season, May to September. The state's scenic beauty, wilderness setting, and wildlife continue to attract and enthrall visitors. Alaska has an opportunity to capitalize on the perception that the state is a safe place to visit and an exotic alternative to traveling abroad. Abundant resources are available for communities and businesses to develop cultural tourism, ecotourism, wildlife viewing, adventure tourism, and sportfishing opportunities.

The cruise industry has had the greatest impact overall. Virtually all visitor growth in the 2000-2001 season is attributable to the cruise sector. Additional growth is expected over the next few years as cruise lines continue to increase their capacity in the market.

Chapter Eight: North to the Future

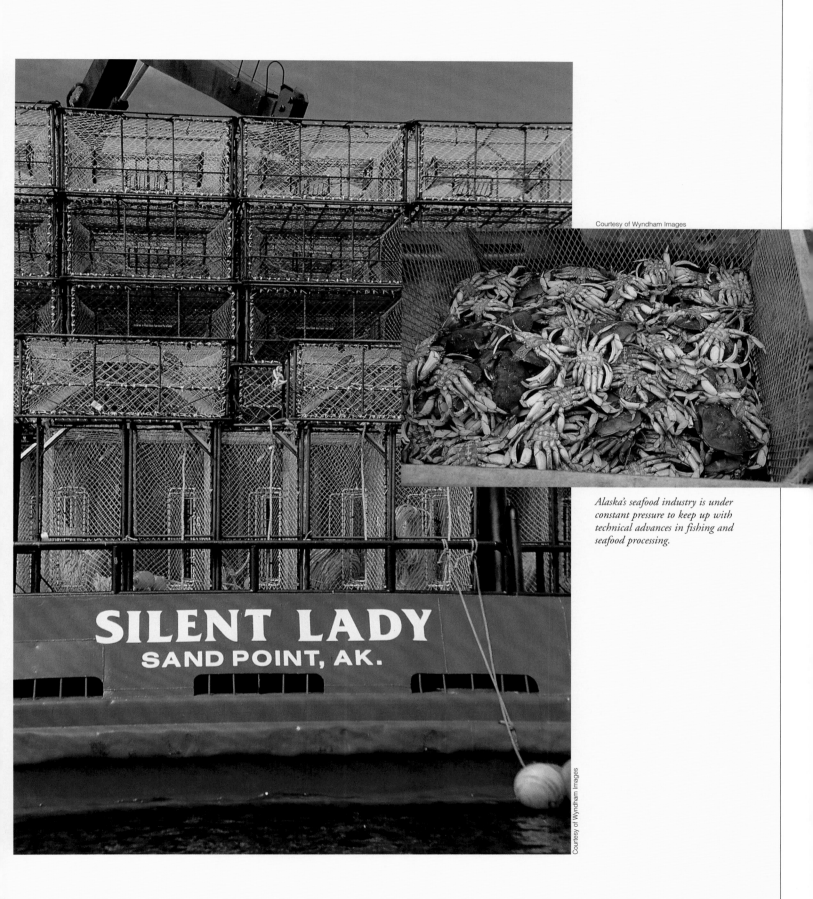

Alaska's seafood industry is under constant pressure to keep up with technical advances in fishing and seafood processing.

Chapter Eight: North to the Future

Everyone enjoys the carnival rides at Anchorage Fur Rendezvous.

As the number of visitors grow, maintaining the quality of the "Alaska experience" at prime attractions is an issue facing the industry. Both the state and the industry recognize this concern and, through long-term planning efforts, are working to improve the visitor infrastructure and develop new attractions.

There are also significant opportunities statewide to expand tourism during the winter season. Winter visitors are drawn by the northern lights, particularly in Fairbanks and the Interior. Aurora viewing is accompanied by other activities, including dog-sled tours, skeet shooting, cross-country and downhill skiing, snow machining, ice skating, ice fishing, and other winter activities.

MINERALS INDUSTRY

With the completion of a mill optimization project at the Red Dog Mine, statewide development expenditures in the minerals industry dropped from $142 million in 2000 to $81 million in 2001. The development of the True North Mine near Fairbanks, the Pogo gold project near Big Delta, and the Kensington/Jualin and Greens Creek Mines near Juneau also helped to cushion the decline. There are several large projects that promise development in the future including the Pogo gold property, the Donlin Creek gold project near Crooked Creek, and continuing discoveries near the Red Dog Mine. Unprecedented increases in the price of platinum, palladium, and tantalum spurred many smaller exploration projects in 2001, from platinum and palladium in Union Bay in Southeast Alaska and near Paxon to Kougarok (tantalum) on the Seward Peninsula. The billion-ton Pebble Copper project west of Iliamna was the subject of renewed interest after detailed geophysical surveys showed that the known copper-gold mineralization may be the tip of the iceberg. In the same area, the million-ounce Shotgun gold prospect was the focus of exploration in 2002.

In short, there are several good exploration targets in advanced stages. The projects' economic impact will improve industry values when they move from exploration to development and production.

There are further reserves in the vicinity of Fort Knox gold mine, north of Fairbanks, that may provide feedstock for the existing mill. The future of the Red Dog zinc-lead-silver mine north of Kotzebue seems assured with huge reserves in the vicinity of the existing mine. Greens Creek Mine near Juneau also has sufficient reserves to last

A cruise ship docks at Juneau.

several years, and the chance of further discoveries in the mine is good, as is also the case at Illinois Creek. Alaska also has substantial known reserves of many other metallic and non-metallic commodities such at tin, beryllium, barite, molybdenum, rare earths, and graphite. Coal resources in the northwest Arctic could amount to four trillion tons, and the heat content of the identified resource matches some of the best thermal coals elsewhere in the world.

Wood Products Industry

Recent years have brought the Alaska forest products industry to its lowest point in fifty years. Three major trends have had negative effects: first, the long-term stagnation of Japan's economy, Alaska's primary export market; second, a substantial decrease in allowable harvest levels in the Tongass National Forest; and third, a decrease in harvest on privately held Native Corporation lands. The only modest growth that has taken place has been with small firms in Southcentral and Interior Alaska, instead of Southeast, the traditional stronghold of the forest products industry.

Despite the gloomy economic outlook for the industry, some positive developments are taking place. The Alaska Wood Technology Center in Ketchikan is currently testing the strength characteristics of Alaska tree species in order to establish Alaska-specific lumber grades. It is thought that these grades will recognize the higher design values of Alaska species that are currently lumped into less advantageous grades. The new grades will increase the values of Alaska lumber and standing timber. Further, in Southeast Alaska, two efforts are underway to realize profit from currently under-utilized parts of the resource. With assistance from the Department of Energy, Sealaska Corporation is investigating the feasibility of a facility that would convert wood waste into ethanol, thus lowering or offsetting the disposal costs and future liability of wood waste from both harvesting and manufacturing. A group of investors is also trying to restart the veneer mill in Ketchikan formerly owned by Gateway Forest Products. If functional, the mill would use lower-grade hemlock and spruce. By trading with other mills for a log supply, this would increase production efficiency for the area's entire industry.

Fireworks at Fur Rondy.

Chapter Eight: North to the Future

Blessed with three instrumented runways, ANC is an all-weather facility on which industry and the travelling public relies.

Snowboarding competition at Fur Rondy.

Chapter Eight: North to the Future

The Raffles Light arrives at the Port of Anchorage. Refined petroleum products are both imported and exported through the Port.

Courtesy of Port of Anchorage; © Greg Martin

Chapter Eight: North to the Future

Central Gathering Facility.

TRANSPORTATION AND LOGISTICS INDUSTRY

The transport of people and products plays a much larger role in Alaska's economy than in much of the rest of the nation. The lack of truck and rail transportation means, in many cases, that resources cannot be developed.

By Air. Air transportation accounts for half of all transportation employment in Alaska compared with less than one-third nationally. In Anchorage, one in ten residents works at a job that is airport related. There are more than 1,100 airstrips and airports in Alaska, almost 10,000 registered aircraft and as many pilots. The state owns or operates 171 gravel-surfaced airports and 43 paved airports as well as numerous seaplane bases. Municipalities own or operate another 20 airports. Ted Stevens Anchorage International, Fairbanks International, Juneau International, and Ketchikan International airports account for most air activity occurring throughout the state.

Anchorage's air-cargo industry is expected to continue to expand by an average of five percent annually over the next five years, mirroring worldwide market trends, but down from the double-digit growth experienced in much of the 1990s. Air-cargo industry analysts report that a glut of cargo airplanes in the world market will likely result in lower freight prices and allow more goods to be shipped cheaply to U.S. markets, using Alaska as a fueling stop. In 2000, Ted Stevens Anchorage International Airport led all U.S. airports in the amount of fuel pumped into cargo planes—more than 700 million gallons. The airport annually ranks among the nation's top ten cargo airports, averaging approximately 520 cargo flights weekly.

By Land. During the closing months of 1999, the U.S. Congress passed a new highway appropriations bill increasing Alaska's highway funding by almost 50 percent. The new legislation provided nearly $100 million annually through 2002 and supports the largest highway construction and maintenance program in Alaska since statehood. Today, 1,487 miles (73 percent) of Alaska's National Highway System roads meet national standards. During fiscal year 2002, the statewide highway budget was approximately $350 million dollars, covering scores of projects ranging from reconstructed roads and bridge replacements to trail safety marking and new construction.

Anchorage is the hub for the Alaska Railroad, with rail access to the ocean ports of Seward and Whittier, and to communities as far north as Fairbanks. In the last several years, the rail system was significantly strengthened through the introduction of year-round container-ship service at the Port of Anchorage and railcar-barge service between Alaska and the continental United States.

By Sea. Waterborne access remains an essential component of economic development in Alaska. In regions of the state unconnected by roads, tug and barge operations provide a vital service to communities depending on barges for most of their supplies and heating oil. The vast majority of Alaskans employed in marine operations work for private companies, including tug and barge operations, chartering, lightering, and warehousing. The Alaska Railroad, in partnership with private companies, provides direct shipping of individual railroad cars aboard mainline barges destined for the Port of Whittier.

According to the State Department of Transportation and Public Facilities, beginning in 2000, Alaska's Marine Highway ferries experienced a shift in marine freight and passenger service, from mainline long-haul service toward point-to-point local service. In Southeast Alaska and Prince William Sound, the addition of smaller, high-speed ferries over the next several years is expected to dramatically alter how people and goods move throughout rural coastal marine areas unconnected by roads.

Alaska's seaports and coastal harbors are principal centers of commerce and crucial links to interior communities. New and improved ports and harbors can reduce the delivery cost of goods and services, increase the frequency of delivery, improve the value of regionally exported resources and products, and improve the productivity, safety, and quality of life for people in a region.

Information in this chapter is courtesy of Alaska Department of Community and Economic Development, the text is comprised largely of excerpts from the Department's publication, *Alaska Economic Performance Report 2002.*

The Port of Anchorage is a hub of a massive multimodal transportation system that connects over 80 percent of the state by truck, train, barge and plane connections.

Courtesy of Port of Anchorage; © Greg Martin

Chapter Eight: North to the Future

Endicott.

Chapter Eight: North to the Future

Courtesy of Wyndham Images

While agriculture does not make a large impact on the state economy, the Mat-Su valley is home to many farms (right). The long hours of sunshine during the summer often do have an interesting effect on produce. Some things tend to get extraordinarily large during the short season. For instance, it is not unusual to see extra-jumbo-sized cabbages or pumpkins growing in the garden. Unfortunately, the rapid growth does not allow sugars to form, so taste may be a problem with some of Alaska's oversized vegetables. Research is being done to find vegetable strains that can grow rapidly and still form sugars to make the vegetables more palatable.

Chapter Eight: North to the Future

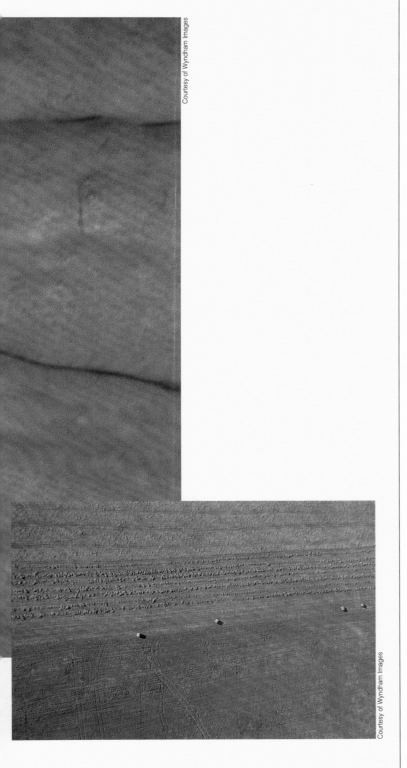

"You'd Never Think That in Alaska There'd Be…"

A Future in Agriculture

Most people would not think of agriculture in the state of Alaska. They certainly didn't over 100 years ago, when Charles C. Georgeson traveled north as a special agent in charge of the United States Agricultural Experiment Stations. In 1898, Georgeson arrived in Sitka and set about the task of finding out if crops and farm animals could survive in the recently acquired lands. The agricultural studies that he set in motion over a century ago are still carried on today at the University of Alaska Fairbanks' Agricultural and Forestry Experiment Station.

Georgeson had a challenging start, but he did establish that crops could survive in the far north, some better than others. He quickly helped to establish several other stations. The final three are the only ones still remaining: the Fairbanks station, which opened in 1906; the Matanuska farm station, which opened in 1915; and the Palmer Research Center, which opened in 1948.

Today, farmed land accounts for only a fraction of one percent of the state. While agriculture is practiced throughout the state, it is typically on a small scale and for local consumption. Commercial agriculture occurs mostly in Southcentral Alaska in the Matanuska-Susitna Valley and Tanana Valley, and also to a small degree in the Copper River Valley, the Kenai Peninsula, Kodiak Island, and the Aleutian Islands.

Alaskan-grown crops and animals account for about 10 percent of what Alaskans consume. But what of the future? What potential is here? Perhaps, with additional research and development, it may no longer be cheaper to import food from outside Alaska—and Charles C. Georgeson's vision for the future will be realized.

A Future in Space

It may be surprising to know that Alaska is a prime location for aerospace activity. It's true! Alaska's location, higher latitudes, wide-open corridors, and developing support infrastructure all contribute to the potential for it to be among the world's leaders in this exciting industry.

Poker Flats. In fact, the Poker Flat Rocket Range has been in use since 1968. It is the world's only non-federal scientific rocket-launching facility owned by a university. Located 30 miles north of

Chapter Eight: North to the Future

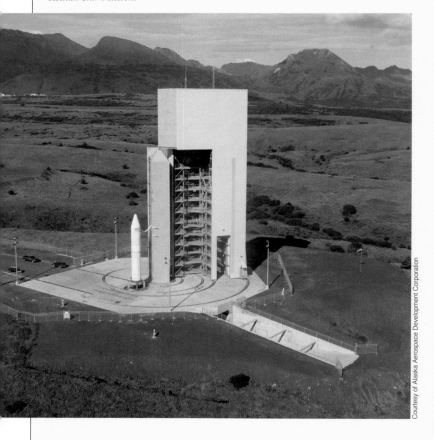

Aerial view of LSS with the Lockheed Athena 1 rocket on the pad for the Kodiak Star Mission.

Fairbanks, the range is under contract to NASA's Wallops Flight Facility. Used by governments, military, and scientists from around the world, it is primarily dedicated to the launch of sub-orbital sounding rockets for the purpose of auroral and atmospheric research. Student rockets are also launched. Recently upgraded, the range also features many scientific instruments designed to study the arctic atmosphere and ionosphere. The University of Alaska Fairbanks is the lead institution for the Alaska Space Grant Program and the research center for the statewide university system. It is one of only 37 institutions in the nation that was eligible for designation as a Space Grant Institution.

Kodiak Launch Complex. *In January 1998, the Alaska Aerospace Development Corporation began building a 3,100 square-foot commercial spaceport at Narrow Cape on Kodiak Island. Its first mission was launched in November of that same year—a sub-orbital vehicle for the U.S. Air Force called "ait-1." A second U.S. Air Force rocket was launched in September 1999 from the Kodiak Launch Complex for the atmospheric interceptor technology program. Climbing to more than 1,000 miles, the "ait-2" was monitored to its final destination off the southern Oregon coast.*

Located 250 miles southwest of Anchorage, Kodiak Island is one of the best locations in the world for polar launch operations, providing a wide launch-corridor and unobstructed downrange flight path. The facility's location and innovative low-cost operations also make it ideal for launching telecommunications, remote sensing, and space-science payloads of up to 8,000 pounds into low earth polar (LEO) and Molniya orbits. The Kodiak Complex is a state-of-the-art launch facility, offering all-weather, in-door processing that is flexible, economical, and adaptable to all current small rocket-launch vehicles. It is the only commercial launch range in the country that is not cooperatively located to a federal facility.

With an eye toward developing this academic discipline for future generations, in the summer of 2000 the Challenger Learning Center opened on the southern Kenai Peninsula, providing space-education programs for school children. Part of a nationwide network of Challenger Centers, it offers distance-learning programs for schools across Alaska and teacher training for specific curriculum units. It also serves as an interactive educational museum and space camp for area students.

Chapter Eight: North to the Future

Kodiak Star payload processing of four satellites.

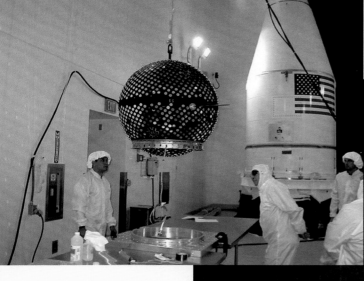

Rocket launch.

Part Two

Prominent Communities of Alaska

Akutan is a fishing community that is located in the eastern Aleutian Islands.

Sitka

From top to bottom:

Downtown Sitka, looking up Lincoln Street toward the Russian Orthodox Cathedral.

Downtown Sitka is framed by the many mountains of Baranof Island.

Sitka Tribe of Alaska's Sheet'ka Kwaan Naa Kahidi community house.

Sitka . . . where rainforest-covered mountains skim the sky and reach down to touch the ocean. One can catch a trophy king salmon while whales breach nearby, stand in the exact spot where the U.S. flag first flew on Alaskan soil, join a raft of otters as they lounge in their snug kelp bed, or be transported back hundreds of years in time while watching Tlingit dancers and storytellers perform before the fire pit in the community house they own and operate.

Sitka is unique among Southeast Alaska's communities—its natural port, located off spectacular Sitka Sound, opens directly onto the Pacific Ocean. This means the mountains are a bit higher, the salmon a bit bigger, and, some say, the people a bit more adventurous than elsewhere along the Inside Passage.

Sitka enjoys many dramatic yearly migrations. Humpback whales spend spring through early winter in Sitka Sound, then swim south to Hawaii to calve. Each fall, Sitka's WhaleFest commemorates the event by inviting scientists to share their latest research. The Sitka Summer Music Festival has, for more than 30 years, presented a June-long series of concerts featuring renowned musicians from around the world. The Sitka symposium on Human Values and the Written Word hosts authors of the caliber of Barbara Kingsolver and Wendell Berry. James Michener wrote *Alaska* at a historic home at Sheldon Jackson College. John "Duke" Wayne used to come up every summer to fish. Charles Kuralt called Sitka "the quintessential small town."

Sitka is the most historic town in all of Alaska. By the middle of the nineteenth century, Sitka was known as the "Paris of the Pacific," the largest and most developed port on the West Coast of North America. Russia's American Empire reached its apex here, leaving a legacy of schools and churches, military installations and a scientific station, boatyards, foundries, sawmills, and a governor's mansion atop Castle Hill. This is the same Castle Hill where the 1867 transfer of Alaska from Russia to the U.S. is re-enacted each October 18th, Alaska Day, as the centerpiece of a weeklong celebration of period-dress balls, contests, and concerts.

From top to bottom:
Gold miner statue outside Sitka's Pioneer Home.
Lighthouse in Sitka.
Naa Kahidi Dancer of the Sitka Tribe of Alaska.

The Tlingit Indian name for Castle Hill is *Noow Tlein* or "Big Fort" and refers to the Tlingit use of the location as an important refuge site. Archeologists inform us that Tlingit people were at the eastern side of Baranof Island 10,000 years ago. Oral narratives from Tlingit elders tell migration stories from the days of the ice ages. Sitka Tlingit people have been renowned for centuries for their artwork, dancing, and storytelling. Today, these traditions are kept alive and proudly shared at the Sheet'ka Kwaan Naa Kahidi Community House.

Sitka is a vibrant, visitor-friendly community with the most diverse economy in Southeast Alaska. With more than 200,000 cruise-ship visitors annually, Sitka appreciates tourism as a foundation of the economy. The town is also a favorite spot for independent travelers. They visit for several days to several weeks, using Sitka as a jumping off point for wilderness expeditions by kayak or aboard the most luxurious yachts. The town has also become a prime spot for conventions, seminars, and workshops as word-of-mouth spreads and Sitka is featured again and again in major national publications and network television.

Sitka offers unrivaled scenic beauty and an unlimited number of outdoor recreational activities. Natural features abound. For instance, an adventurer may take a hike on the trail to Mt. Edgecumbe, a dormant volcano that could be Mt. Fuji's twin, located across from Sitka on Kruzof Island. Another turn off the trail leads to a white-sand beach with waves so impressive that intrepid surfers don wetsuits to ride them.

Auto- and van-accessible Harbor Mountain Road leads to stunning alpine views over the island-dotted waters of northern Sitka Sound. A spectacular trail begins at the top and follows alpine ridges all the way back to town.

But this is by no means just a "tourist town"—employers in health care, government, education, and the largest commercial fishing fleet in Southeast Alaska balance Sitka's economy. Sitka is a main Southeast center for salmon, halibut, black cod, and herring, as well as the new dive fisheries. The U.S. Coast Guard maintains both an Air Station and an Aides to Navigation team in Sitka.

While timber harvests have been greatly reduced, state-of-the-art helicopter logging and other new techniques are allowing more careful use of forest resources. Industrial space and abundant, affordable water and electrical power are available at the new Sawmill Cove Industrial Park.

From top to bottom:

A fishing boat returns to Sitka with a bountiful harvest.

Visitors to the Alaska Raptor Center meet Volta.

Halibut are plentiful for both commercial and sport fishermen.

From top to bottom:

A totem pole stands proudly in the Sitka National Historical Park.

The New Archangel Dancers of Sitka capture a time when Russia ruled Alaska from Sitka.

Whale-watching aboard an Allen Marine Tours' boat.

Photo credits from left to right

Courtesy of tSitka Convention and Visitors Bureau; Photo by Clark James Mishler

Courtesy of tSitka Convention and Visitors Bureau; Photo by John Hyde

Courtesy of tSitka Convention and Visitors Bureau; Photo by Clark James Mishler

Courtesy of tSitka Convention and Visitors Bureau; Photo by Clark James Mishler

From top to bottom:

The commercial fishing fleet is in town.

Overlooking the Pacific Ocean and Mt. Edgecumbe in Sitka.

Kayakers in Sitka waters.

Photo credits from left to right:
Courtesy of tSitka Convention and Visitors Bureau;
Photo by John Hyde, Wild Things Photography
Courtesy of tSitka Convention and Visitors Bureau;
Photo by John Hyde, Wild Things Photography
Courtesy of tSitka Convention and Visitors Bureau;
Photo by John Hyde, Wild Things Photography
Courtesy of tSitka Convention and Visitors Bureau;
Photo by Clark James Mishler
Courtesy of tSitka Convention and Visitors Bureau;
Photo by Clark James Mishler
Background photo courtesy of Wyndham Images

Sitka is the site of the private, four-year Sheldon Jackson College, as well as a branch of the University of Alaska. There is a state-run boarding high school, an alternative high school, and an array of vocational programs. Sitka's excellent public K-12 schools consistently garner awards and trophies for programs ranging from jazz band to swimming to drama and debate. Graduates often move on to top colleges and universities.

Although Sitka may be remote, Sitkans do not want for many "Big City" services. Sitkans participate fully in the World Wide Web with Internet services available, including broadband, and generous satellite capacity provides for wireless phones and competitive long-distance rates. Sitka also has surprises for a town of 8,500—a daily newspaper, for instance, and an award-winning public radio station that are cornerstones of local and regional communications.

Two hospitals, plus a wide variety of community services, make senior or disabled Sitkans and visitors alike feel comfortable. Many local attractions and several trails are barrier-free. There is even a remote U.S. Forest Service site that provides ADA accessibility to the cabin, outhouse, and a deck that hangs just a fishing line's length above the lake.

More than a score of churches provide religious services for most denominations. Dining ranges from steak-and-seafood clubs to neighborhood snack bars. Lodging runs the gamut from full-service hotels to quaint bed-and-breakfast establishments offering ocean views and other Sitka-style amenities.

Sitka enjoys regular and reliable barge and ferry service. Sitka Rocky Gutierrez Airport, a modern facility with full services such as rental cars and air-taxi service, provides year-round daily jet connections to Juneau, Anchorage, and Seattle.

Famous for small-town friendliness and wonderful quality of life, Sitka brings together a diverse population, about 25 percent of whom are Alaska Native. New residents are welcomed from all 50 states and quickly make themselves a part of the community. Naturally, Sitka is a prime retirement spot, with seniors from both Alaska and the Lower 48 deciding to spend their golden years in a place that ranks among the most lush and beautiful landscapes in the world.

Unparalleled beauty, limitless outdoor adventure, a rich culture and history, a thriving economy in a small town known for its friendliness—all can be found in Sitka, Alaska.

Will Swagel

From top to bottom:

The Sitka Summer Music Festival at Harrigan Centennial Hall in Sitka.

Naa Kahidi Dancers of the Sitka Tribe of Alaska.

The City and Borough of Juneau

From top to bottom:

In June, more than 1,000 dancers clad in formal regalia ushered in Celebration 2002, a biennial event held in Juneau honoring Native heritage and traditions.

Two hikers enjoy the summertime panorama before them, with blazing fireweed in the foreground and Mendenhall Glacier in the background.

Photo credits from left to right:
Photo © David Job
Photo © 1994 David J. Job
Photo © Patrick McGonegal
Photo © Patrick McGonegal
Photo © Patrick McGonegal
Background photo courtesy of Wyndham Images

Nestled between mountains and sea, the City and Borough of Juneau is located in the Panhandle of Southeast Alaska, 900 air miles north of Seattle and 600 air miles southeast of Anchorage. Within the 3,000 square miles of the City and Borough of Juneau—including the spectacular Juneau Icefield—live over 30,000 people. Juneau has a Tlingit history with a strong historical influence from the early prospectors and boomtown that grew around gold-mining operations. One of the most beautiful state capitals in the nation, the economy is based on government, tourism, mining, and fishing. Like many other Alaskan towns, residents of Juneau share their home with bears, eagles, mountain goats, salmon, and whales.

Juneau, Alaska's capital for nearly 100 years, is committed to being the finest state capital in the nation and continues to work hard to be worthy of that honor. As Alaska's Capital, Juneau annually hosts State Legislators and their staff during the legislative session between January and May. Juneau is one of the most technologically connected cities in the nation, with satellite and fiber optic links to the rest of the world. Alaskans enjoy some of the best access to government and legislative information and deliberations through the Internet, television, and streaming online video. Jets into the Capital City now carry state-of-the-art GPS navigation providing one of the best approach and departure systems in the country.

Still strongly dependent on federal, state, and local government jobs, Juneau is by no means simply a government town. Its diverse and growing economy includes contributions from tourism, fishing, mining, construction, and manufacturing. Juneau's entrepreneurial spirit has produced an award-winning brewery, smoked salmon with a national reputation, and roasted coffee that is the best in the region, to mention only a few. During the summer months, the tourism industry injects over $90 million into the local economy, providing almost 2,000 seasonal jobs. More than 730,000 cruise-ship passengers visit Juneau annually.

From top to bottom:

The spirit of adventure may draw some to the singular sport of ice climbing—on a glacier!

In Juneau, even the least hardy stroller can take a short drive and then quite literally walk up to a glacier and contemplate the grandeur and awesome power of it.

Mendenhall Glacier in winter.

From top to bottom:

A city of both mountains and sea, Juneau is situated in a dramatic landscape carved by glaciers.

Juneau's colorful downtown is always bustling, but is especially alive during the summer months.

Whether it is playing in the Gold Medal Basketball Tournament or shopping in the local stores, Alaskans from neighboring Southeast towns come in and out of Juneau regularly. Often, while in Juneau, they indulge in a weekend of skiing, take in an evening at the symphony, or attend a live performance of a good play. Juneau offers its residents and visitors a wide variety of recreational and cultural opportunities.

Among the many opportunities for outdoor adventure available in Juneau are the over 100 miles of groomed trails for hiking and biking. During the winter months, these same trails are used for snowshoeing and cross-country skiing. The local ski area, Eaglecrest, is owned and operated by the City and Borough of Juneau and provides a safe, family-oriented winter recreation activity.

Bartlett Regional Hospital provides a modern range of services not available in the average community of over 30,000 people. Specialized care includes cardio-pulmonary rehabilitation, a family birth center, a mental-health unit, and an on-site MRI. Bartlett provides services throughout all of Southeast Alaska using telemedicine technology linking rural locations to Bartlett and to hospitals in the Lower 48.

Incredibly, the City and Borough of Juneau encompasses 3,248 square miles in total, including 928 miles of ice cap and 704 square miles of water. Of that amount, a total of 264 square miles are urbanized. Juneau has grown along both banks of the Gastineau Channel on Douglas Island and the mainland, filling the valleys of Lemon Creek and Mendenhall, which were carved by glaciers. More than two-thirds of the population lives in these areas.

From top to bottom:

A seaplane streams along Gastineau Channel.

A cruise ship glows against the dark waters of Gastineau Channel.

Juneau's waterfront docks are silhouetted against the shining waters of Gastineau Channel.

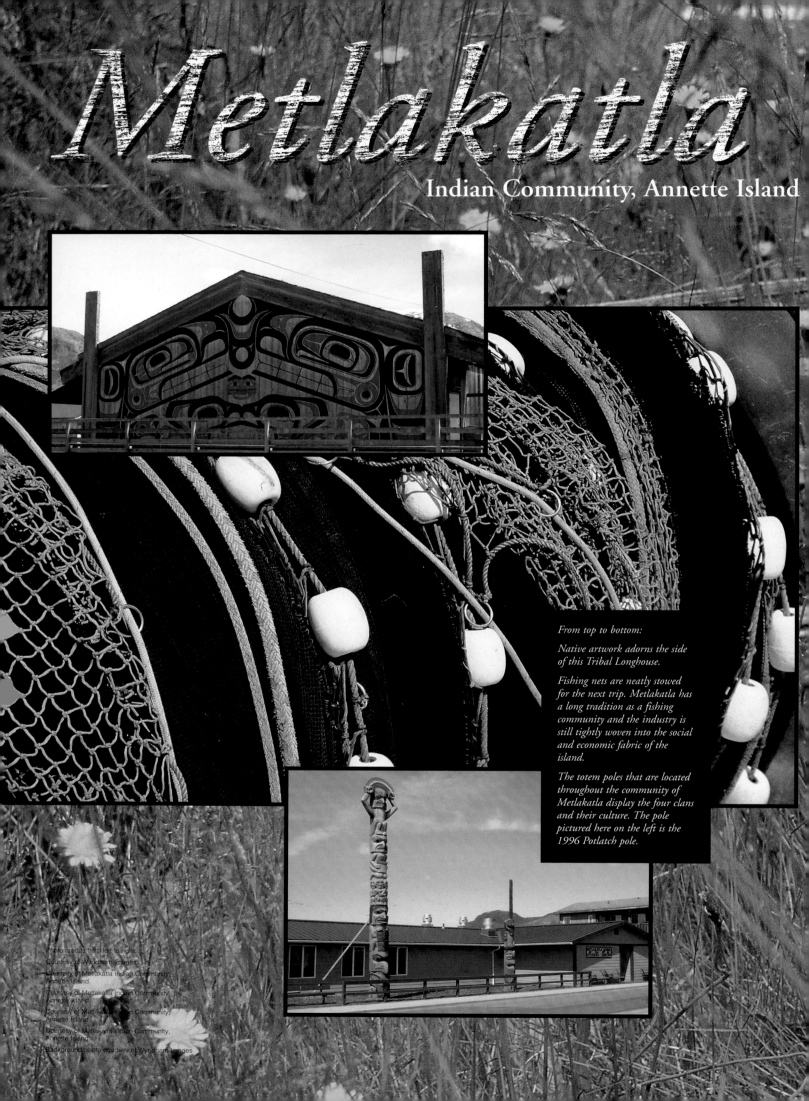

Metlakatla
Indian Community, Annette Island

From top to bottom:

Native artwork adorns the side of this Tribal Longhouse.

Fishing nets are neatly stowed for the next trip. Metlakatla has a long tradition as a fishing community and the industry is still tightly woven into the social and economic fabric of the island.

The totem poles that are located throughout the community of Metlakatla display the four clans and their culture. The pole pictured here on the left is the 1996 Potlatch pole.

At the very tip of the Southeast panhandle of Alaska, about 15 miles south of Ketchikan and 700 miles from Seattle, sits a green and purple island called Annette. At times serene during the warm, sunny days of summer, at times forbidding when the autumn winds blow in October and November, Annette Island is, at any time of year, incredibly beautiful.

Annette Island is the only Indian reservation in the State of Alaska. In 1887, a total of 863 Tsimshean tribe members relocated from British Columbia, Canada and settled on the island. They established the town of Metlakatla (a Tsimshean word meaning, "calm channel"), which is the only community on Annette Island. In 1891, the U.S. Congress established the Annette Islands as a Reservation set aside for the Tsimshean Indians and other Alaskan Natives who may wish to join in membership. Today, there are 2,094 tribally enrolled members, with approximately 1,300 currently living in Metlakatla.

Annette Island is 21 miles long, 14 miles wide, and encompasses 87,000 acres of mountainous, tree-filled terrain. With exclusive water rights, the Reservation of Annette Islands is 122,000 acres overall. Flying over the island, one can spot at least a dozen large freshwater lakes and hundreds of small ponds. Everyone's favorite mountain is "Purple Mountain" (it really is purple!), which has a picturesque waterfall facing the community of Metlakatla. This waterfall has been a main attraction for photographers over the years. Another unique attraction is "Yellow Hill" (again, an accurate description). A climb to the top of Yellow Hill will yield a 360-degree view of at least one-third of Annette Island.

Metlakatla is an eager and ambitious community that is blessed with an abundance of natural resources, and Annette Island's geological structure is a boon to investors and developers. Annette benefits from a deep-water harbor and docking facility that supports large vessels up to 500 feet. The island boasts the only flat-based land in Southeast Alaska that can support industrial and recreational uses. Of the large, flat spaces available, one area contains a 10,000-foot-long airstrip that is functional and available for those economic ventures requiring airfreight transport. The use of land for residential and business purposes is established and controlled by the Metlakatla Tribal Council.

The exclusive fishing rights of the Annette Islands were established in 1916 and are still enjoyed by island residents today. Metlakatla has always been a fishing community and is still home to a large number of independent fishermen. The community owns and operates a fish hatchery and a cold-storage plant that processes the local catches of all fish products and sells the finished product globally. The cold-storage facility flash freezes salmon and other species of commercially viable seafood.

From top to bottom:

Evidence of Metlakatla's strong sense of Native heritage is everywhere. Beautiful totem poles stand proudly throughout the community. Here stands the Frank R. Hayward Memorial Totem pole.

Bald Ridge Mine site.

From top to bottom:

Silhouetted by Purple Mountain, Metlakatla is bursting with potential. With the right combination of natural beauty, natural resources, and the soon-to-be-completed easy access to Ketchikan via Waldon Point Road, Metlakatla promises great opportunity to the right developers.

Metlakatla Indian Community owns and operates a cold-storage plant that flash freezes seafood, such as this fresh-caught and locally processed shrimp, for markets in the Lower 48 and around the world.

All inset photos courtesy of Metlakatla Indian Community, Annette Island
Background photo courtesy of Wyndham Images

Metlakatla has embarked upon some major projects to sustain and diversify its economic base. The first example is Bald Ridge Mine, a 200-million-ton rock reserve that is in the final stages of development. The product is a high-quality rock used for highway projects and for the construction of marine jetties. The rock can also be cut for decorative and functional uses such as kitchen countertops. When fully operational, the mine will provide jobs for local residents and generate three to five million tons of rock annually. Metlakatla is in a prime position to sell high-grade aggregate rock to the entire Pacific Rim. Other projects in development include a bottled-water plant in the construction phase and a native-seed nursery in the planning phase.

There are many Native artists, including carvers and jewelry makers, whose works are gaining recognition throughout Alaska and beyond. Many of these artists display and sell their work to visitors who enjoy and value Native art at the local Artist's Village, a facility built by the community. Metlakatla takes pride in its Native heritage—there are five Native dance groups, and totem poles are located throughout the community.

In 1997, Metlakatla began construction of Waldon Point Road with the assistance of the Army Corps of Engineers, the Federal Highway Administration, and Alaska Department of Transportation. The 14.7-mile-long highway will more effectively connect Metlakatla to the City of Ketchikan. When completed (projected to be 2007), the road will extend from Metlakatla to Annette Bay on the other side of Annette Island and will terminate with a short ferry ride to Saxman and Ketchikan. A new 180-foot ferry is also under construction and will provide daily ferry-shuttle service between the northern ferry terminal on Annette Island and the City of Ketchikan beginning in 2004. The shuttle is a short 20-minute ride with 150-passenger and 19-vehicle capability. Residents of Metlakatla will soon have easier access to health services, the Ketchikan branch of University of Alaska, shopping, and recreational facilities. The transport of goods and services will be facilitated and tourism will grow due to increased access to the island. Tribal members will have fresh opportunities to build retail, food service, and recreational facilities that will attract new visitors and expand the island's economic base.

The Council of Metlakatla has declared its intent to provide the opportunity for large and small businesses to establish economic development within the Annette Islands Reserve on a long-term basis. The Council also invites visitors to view beautiful Annette Island, experience its rich heritage, examine its bountiful resource land base, and chat with the friendly Native peoples of the Metlakatla Indian Community. The future expansion of Metlakatla's economic development is virtually limitless, as its beauty is boundless.

From top to bottom:

The truly stunning views of and from Annette Island speak for themselves. This vista is of Tamgas Harbor as seen from Yellow Hill.

In 1997, Metlakatla began the construction of Waldon Point Road, a 14.7-mile-long highway that will more effectively connect Metlakatla to the City of Ketchikan.

Fishing has always been at the heart of Metlakatla's economy. There are still a large number of independent fishermen based in the community. Here, fishermen are readying the local seine boat Migrator.

The City of Wrangell

From left to right:

This interior Tlingit housepost detail is from Chief Shakes Tribal House in Wrangell.

Wrangell's sheltered deep-water harbors are a boon to the city's fishing and seafood-processing industries.

Beachcombers find something quite different at Petroglyph Beach State Park. The beach is full of unusual rock carvings that are thousands of years old.

Located at the mouth of the Stikine River in the heart of the Tongass National Forest, the city of Wrangell is one of Southeast Alaska's gems. Wrangell is a true Alaskan town, picturesque and accessible to some of the premier recreational areas in North America. A friendly community with a rich and varied cultural heritage, Wrangell plays an important role in the rich history of Southeast Alaska. There is a strong sense of community values that sets Wrangell apart from larger urban areas.

Wrangellites are open, warm, and ready to accept any challenge. Local citizens are always willing to adapt to economic conditions and are quick to take advantage of economic opportunities. Planning for a prosperous economic future, Wrangell has erected the critical infrastructure necessary to support future development. Wrangell boasts two of the best, sheltered, deep-water harbors in the region and plans are on the immediate horizon to construct a third harbor. Improvements to harbors and facilities have been made to further support Wrangell's seafood-processing industry and commercial fishermen. Local timber processors also have a good relationship with the City, which helps to assure economic diversity and resources for continued economic growth. Among the City's holdings are key industrial properties adjacent to the deepwater docks, downtown, and near the airport, which are primed for development.

Wrangell is located equidistant between Seattle, Washington and Anchorage, Alaska and is only 150 miles south of Juneau, Alaska. Visitors to Wrangell typically arrive from either port city by Alaska Airlines jet service or via a leisurely boat ride through the scenic Inside Passage on the Alaska Marine Highway ferry system.

Visitors to Wrangell enjoy many and varied recreational opportunities. With Wrangell's proximity to wilderness areas, outdoor enthusiasts can go hiking, camping, boating, and wildlife-viewing independently or with a locally guided outfitter and guide. Many unique area attractions offer spectacular scenery such as Shakes Glacier on the Stikine River, or Anan Bear and Wildlife Observatory with up-close viewing of bears in their natural habitat. Local "must see" attractions include Petroglyph Beach State Historical Park, full of rock carvings thousands of years old; Chief Shakes Island, with its clan house and totems and Muskeg Meadows; and a regulation nine-hole golf course!

From top to bottom:
This bear has just caught his lunch!.
A purse seiner at work.
This lucky fisherman doesn't have to tell a "fish story."

The City of Seward

From left to right:

A young fisherman just can't wait for dinner.

A young Seward resident displays the catch of the day.

A cruise ship sails near Seward.

The Alaska Railroad.

The perfect choice . . .

Nestled between emerald-green mountains and a deep-blue glacial fjord is the historic port that helped build the great state of Alaska. Today, the opportunities for business development in Seward are as brilliant as its setting and far exceed those of its past. If one is looking to open or expand operations in Alaska, Seward is the perfect choice for the future. Seward is the only year-round ice-free seaport in Alaska, and is also served by road, air, and rail. This makes it the most viable import/export hub in the state. This transportation advantage provides a direct link to Alaska's resource-rich Interior, its two largest cities, and almost two thirds of the state's population. Seward is the home of Alaska's finest marine maintenance facilities, and local businesses provide all of the necessary support services for ocean-going vessels. This same thriving community is only a stone's throw away from a wilderness paradise of untouched forests, sparkling rivers, pristine lakes, and shimmering bays. This outdoor utopia offers seemingly unlimited opportunities for hiking, fishing, and so much more. Offering all the conveniences of modern life including high-speed internet access, schools, a library, a museum, galleries, restaurants, and year-round recreation, Seward is not only a great place for business, it is a great place to live.

International investors and the state of Alaska recognize the future potential of the area. The city already sustains thriving industries in shipping, tourism, and commercial fishing and processing, proving the efficiency and effectiveness of its infrastructure as well as the capabilities of its civic leaders. The Alaska SeaLife Center, University of Alaska, and the Qutekcak Shellfish Hatchery provide world-class marine research, as well as wildlife rehabilitation and education. The Alaska Vocational Technical Center in Seward is dedicated to providing a well-trained labor force for many industries. State, borough, and city tax structures are also favorable to business.

There is no better place to build a quality life—the future is bursting with promise in Seward.

All Photos: Courtesy of City of Seward; Photo by Alaskan Memories Photos

From left to right:
Fourth of July Celebration.
Seward Marine Industrial Center.
Seward Halibut Derby.

Part Three

NETWORKS

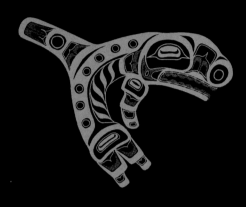

Mt. Huntington.

Golden Valley Electric Association

"All in a day's work," an 8-foot-tall ice sculpture by world-class ice sculptors Vladimir Zhikhartsev and Nadya Fedotova, celebrates the dedication of GVEA's linemen to provide safe, reliable power to over 90,000 Interior residents and businesses.

COOPERATIVE SPIRIT

Golden Valley Electric Association (GVEA) took shape when a small group of people became interested in bringing electric service to rural areas and furthering the agricultural industry in Interior Alaska. That was 1946, and since then, GVEA has grown to serve approximately 90,000 Interior residents in the Fairbanks, Delta, Nenana, Healy, and Cantwell areas at over 39,000 service locations.

Interior Alaska conditions present unique challenges as temperatures can range from 90 degrees to minus 50 degrees, but Golden Valley boasts 99.99 percent system reliability.

As a cooperative, Golden Valley is owned by the members it serves. Its members continually demonstrate their support for the co-op by reelecting their board representatives; its average board member has 13 years of experience.

INNOVATION

Improvements in technology have made it possible to overhaul GVEA's mapping system. The old system involved manually updating three separate maps and entailed adding new subdivisions, poles, and power lines by hand. Currently, GVEA is designing an Intranet mapping system that will consolidate these hard copies into one electronic mapping system. With system entry at their fingertips, employees will be able to view Fairbanks and Nenana recording district plats and GVEA's electrical system; they will also have better access to customer account and billing information, which will improve customer service. Employees will be able to locate poles, meters, and other electrical devices. The system also provides access to U.S. Geological Survey topographical information about terrain that can be important to line design. Providing this information at the desktop level will enable employees to provide members with real-time information and better quality service.

RELIABILITY

Although Golden Valley is proud of its 99.99 percent reliability, the utility is always looking for ways to meet the future needs of its members and improve the service it provides.

One initiative is GVEA's Battery Energy Storage System (BESS). When complete, the BESS will provide 27 megawatts (MW) of power for 15 minutes, although it can provide up to 40 MW for less time if necessary. GVEA anticipates up to a 68 percent reduction in power-supply types of outages when the battery is available. Fifteen minutes is long enough for the co-op to start up and bring local generation online. Other benefits of this unique project include reduced air emissions through reduced spinning reserve operations and improved power quality. Being able to produce 40 MW will

When complete in 2003, GVEA's Battery Energy Storage System (BESS) will be the most powerful battery in the world. It will increase system reliability by providing 27 megawatts of power for 15 minutes.

To provide the best customer service, GVEA's automated mapping system provides real-time information on plats, pole, and meter locations and topographical information.

make the BESS the most powerful battery energy storage system in the world in terms of megawatt output.

To continue to improve reliability even further, GVEA upgraded its SCADA system. SCADA stands for Supervisory Control and Data Acquisition and it's the system the dispatchers use to monitor power supply 24 hours a day. This $800,000 upgrade became necessary because it was hard to find replacement parts for the old system, which dates back to the mid-1980s. This three-year project is now fully operational. The new system works to adjust GVEA's generating units to meet minute-by-minute power requirements. It can also remotely control, monitor, and record system information from Cantwell to Delta.

PLANNING FOR THE FUTURE
Recent growth trends have seen an average of 1,150 new connections to GVEA's system each year. While the majority of these new services are residential, the utility has seen significant load growth in the commercial sector from the construction of the Home Depot to expansions at the Fairbanks International Airport to renovations at local

The 97-mile Northern Intertie, due to be completed in 2003, will bring additional reliable, low-cost energy from Anchorage and Healy to Fairbanks.

Commitment to community is a cooperative principle that GVEA takes to heart. Every year employees donate thousands of hours in the community. Pictured are GVEA employees staffing a Kids Voting precinct.

military bases. This trend of steady load growth is expected to continue with other major projects like the Missile Defense System and future mining opportunities.

Golden Valley is preparing now for the future. While GVEA continues to be optimistic about the future of the 50-megawatt Healy Clean Coal Plant, GVEA needs to start the process now so the generation is available when needed. Evaluating options, planning, obtaining permits and construction take several years. GVEA is expanding its North Pole Power Plant to include a new 57-megawatt gas-combustion turbine that will burn Naphtha, a fuel that produces very low sulfur emissions. GVEA is evaluating wind generation as well. It is Golden Valley Electric Association's vision that by the year 2050, its dependence on fossil fuel will be greatly reduced and the majority of its generation will come from renewable resources. Fifty years may seem like a long way off, but GVEA will be there—serving Alaskans.

From pickups to bucket trucks, and from snow machines to track vehicles, Golden Valley has the fleet of vehicles necessary for all-season response and maintenance.v

GVEA's extensive infrastructure ensures stability and reliability. GVEA is dedicated to building and maintaining systems its members can rely on.

CROWLEY

Crowley has been providing unique solutions to Alaska's logistics and marine transportation challenges for a half century. From Anchorage to the North Slope, the Aleutians to remote river villages near the Canadian border, Crowley has played an important role in Alaska's business development and in protecting its environment.

Crowley entered the Alaska market in 1953 when a Crowley company pioneered the use of barges to transport railcars between Ketchikan, Alaska, and Prince Rupert, British Columbia.

A few years later, Crowley began supplying the Distant Early Warning (DEW) Line radar installations for the U.S. Air Force, including sites on the Aleutian Chain and across the northern coast into Canada. It was the first penetration of the Arctic by commercial tug and barge service. Timing was critical because of the ice pack that remains close to shore near Point Barrow except for about six weeks during the summer.

Crowley provides tanker escort and docking services in Valdez and Prince William Sound for the Alyeska Pipeline Service Company using some of the most technologically advanced and powerful tugboats in the world.

This accomplishment was a prelude to the development of the North Slope in later years.

When oil was discovered in Cook Inlet, first in the Swanson River Field onshore in 1957 and later offshore at McArthur River and other locations, oil-industry officials called on Crowley to help tame the treacherous waters of the inlet. Huge tidal variations and 12-knot currents made a difficult chore of setting platforms without the high-horsepower tugs of today, and no marine support structure was available. Crowley responded to both problems by pioneering a technique of rafting tugs together to achieve the necessary horsepower, by establishing a company called Rig Tenders and

Crowley supports the oil industry on the North Slope, providing heavy hauling, and ice road and ice island construction with its environmentally friendly CATCO all-terrain vehicles.

In the mid-1980s, Crowley launched a service to transport, store, and sell petroleum products throughout all of Alaska's coastline and major Western Alaska river systems.

Dock near Kenai to furnish supply and crew-boat services, and by building six ice-strengthened tug supply boats.

In 1958, Crowley became the first company to offer common-carrier transportation of cargo in containers from the Lower 48 to Alaska. For many years, Crowley barges delivered containerized and general cargo to ports throughout the state, and the company continues to provide contract tug and barge services in support of infrastructure and resource development projects.

In 1963, Crowley commenced regular rail-barge operations, known as the Alaska Hydro-Train, for the Alaska Railroad. This service involved transporting rail cars by barge from Seattle to the Alaska Railroad terminal in Whittier.

When oil was discovered at Prudhoe Bay, the oil industry turned to Crowley. Beginning in 1968, utilizing the earlier pioneering experience in the Arctic, Crowley began the summer sealifts to Prudhoe Bay. Since then, 334 barges carrying nearly 1.3 million tons of cargo have been successfully delivered to the North Slope, including modules the size of ten-story buildings and weighing nearly 6,000 tons.

In 1975, the Crowley sealift flotilla faced the worst Arctic ice conditions of the century. In fleet size, it was the largest sealift in the project's history with 71 vessels amassed to carry 154,520 tons of cargo, including 179 modules reaching as tall as nine stories and weighing up to 1,300 tons each. Vessels stood by for nearly two months waiting for the ice to retreat. Finally, in late September, the ice floe moved back and Crowley's tugs and barges lined up for the slow and arduous haul to Prudhoe Bay. When the ice closed again, it took as many as four tugs to push the barges, one at a time, through the ice. In 2001, Crowley transported the largest modules ever made in Alaska from Anchorage to BP Explorations, Inc.'s Northstar Island and oil field on the North Slope.

Crowley tugs, barges, cranes, and personnel have continued to support North Slope oilfield development and the protection of its environment. Crowley is the marine contractor for Alaska Clean Seas; an oil industry spill response cooperative funded by North Slope producers and the Alyeska Pipeline Service Company.

Crowley provides additional logistics support for oilfield development during the winter on the North Slope with CATCO all-terrain vehicles. These heavy-lift overland transport units have large bag tires designed to work on the frozen tundra without damaging the delicate Arctic ecosystem. In the winter, Crowley employees use these vehicles along with drills and pumps to make ice roads and ice islands for oil exploration. When the weather warms, the ice melts and there is no trace that man was ever there.

Crowley has pioneered solutions to transport and deliver the largest modules ever made in Alaska to the North Slope, and has overcome the toughest waters, weather, and terrain the state has to offer.

Crowley used its technical expertise and specialized equipment to move and set down this drilling rig and platform in the treacherous waters of Cook Inlet.

At the southern terminus of the Trans-Alaska Pipeline, Crowley provides tanker escort and docking services in Valdez harbor for the Alyeska Pipeline Service Company using some of the most technologically advanced and powerful tugboats in the world. During tanker escorts, Crowley tugs are tethered to, or shadow, tankers in the event braking or steering assistance is needed. In 2001, Crowley tugs in Valdez stopped a tanker from colliding with a fishing boat and its nets that had been set across the Valdez Narrows shipping channel. The system worked and waters of Prince William Sound were protected.

Crowley has provided this tanker assist and escort service since 1977, and has positioned other vessels in the area to provide a comprehensive spill prevention and response capability to Alyeska and its member companies.

In the mid-1980s, Crowley launched a service to transport, store, and sell petroleum products throughout all of Alaska's coastline and major Western Alaska river systems. Operating from tank farms in Nome and Kotzebue, the company provides direct delivery of bulk fuels and packaged petroleum products to more than 100 coastal and river villages throughout Alaska. During the warmer months, line-haul barges replenish tank farms and smaller lighterage barges carry fuel to remote villages, often beaching where no docks exist. Crowley also positions "fueler" barges in Bristol Bay for commercial fishing boats during the herring and salmon seasons.

Crowley's Alaska fuel business is being expanded in 2003 with the construction of a tank farm in Bethel. Built in cooperation with the Bethel Native Corporation, the facility will support commercial, aviation, government, and individual fuel consumers in Bethel and outlying villages in the Yukon Kuskokwim region.

Throughout the last half-century, unique expertise and equipment have propelled Crowley into its position as a leader of quality, reliable, and environmentally sound services. Whether its over-the-tundra transportation, supplying fuel in remote areas, providing project cargo services, tanker escort and assist services, or protecting the environment, people who know Crowley, count on Crowley's knowledgeable people to get the job done.

The 1983 sealift was comprised of 26 barges, 14 tugs, and accompanying vessels. Here, the fleet has just arrived in the icy waters of the Beaufort Sea.

Alaska Marine Highway System

M/V Tustumena *operating off of Kodiak Island.*

Celebrating its 40th anniversary in 2003, the ". . . safe, reliable, and efficient transportation of people, goods, and vehicles among Alaska communities, Canada, and the Lower 48," continues today as the mission of Alaska's Marine Highway.

The Marine Highway, a National Scenic Byway, carries vehicles and passengers between coastal communities that, for the most part, have no other road access. Routes run from Bellingham, Washington and Prince Rupert, British Columbia, Canada, through the Inside Passage, Prince William Sound, the Kenai Peninsula, Kodiak Island, and out the Aleutian Chain. The newest ship, the ocean-going M/V *Kennicott*, crosses the Gulf of Alaska routinely, offering a leisurely alternative to the long-haul, overland 1,500-mile Alaska Highway and providing a connection between the System's southeast and south-central routes. It is an "express lane" for commercial shippers and summer vacationers to the state's urban centers and primary road system.

The System is about to embark on a major change with the introduction of its "fast vehicle ferries." High-speed aluminum catamarans, capable of sustained speeds of 32 knots and built to the strictest international safety standards, are under construction. The first high-speed passenger and vehicle ferry to be built in the United States, the M/V *Fairweather* will come into service in 2004. It will make the trip between Sitka and Juneau in a little over four hours—less than half its current running time. The M/V *Chenega* will follow to provide service in Prince William Sound between Cordova, Valdez, and Whittier.

While plying some of the world's most challenging waters, the Alaska Marine Highway continues to maintain an outstanding safety record. Captain George Capacci, general manager, notes, "Safety is not a part-time job. Our mission emphasizes a safe marine transportation system. We take those words seriously throughout the organization every day as we carry on the proud, 40-year tradition of Alaska's Marine Highway."

A view of Mendenhall Glacier from the ferry in Auke Bay.

GCI

The far reaches of Alaska present a telecommunications challenge unique to the state and unlike any other in the U.S. Where there are fewer than 14,500 road miles in a state a fifth the size of the Lower 48 states combined, much of Alaska's regions remain relatively isolated from each other. However, this isolation is quickly becoming a thing of the past through the efforts of the state's leading telecommunication provider, GCI. GCI provides the communication links that are bringing Alaskans closer together and allowing these regions to do business within the state and around the world.

Alaskans spend more per capita on telecommunication services than the residents of any other state and have twice the average number of computers online per capita. In addition, Alaska has one of the youngest populations in the nation and is one of the fastest growing market areas in the U.S. These factors create a great demand for a full array of integrated communication services.

GCI was founded in 1979 by Alaskan entrepreneurs Robert Walp and Ron Duncan, GCI's first and current presidents respectively. The company introduced competition to the Alaska telecommunication market, and since that time, has grown to 1,200 employees and is one of the state's largest private employers.

GCI has taken full advantage of the Telecommunications Act of 1996 by combining all the assets necessary to deliver services that customers want. GCI is the only player in Alaska to have successfully combined all the technologies needed to deliver voice, data, and video services.

GCI is a regional integrated communications provider (ICP) serving the Alaska market. It has an established statewide long-distance business with a 45 percent market share. It provides facilities-based, competitive local exchange services with direct access to 65 percent of the state's telephone lines. It owns and operates cable television services in 18 of the state's largest communities, and is the region's largest Internet service provider on a retail and wholesale basis. In addition, GCI holds statewide narrow-band and broadband wireless licenses and uses its own fiber-optic, metropolitan-area networks and satellite-transmission facilities.

GCI owns and operates facilities throughout the state, including route-diverse fiber-optic cables connecting Alaska with the contiguous United States. It employs an array of transmission media including fiber-optic cable, satellite and hybrid fiber-coaxial cable in its 220-plus points of presence in Alaska and the rest of the U.S.

GCI continues to promote competition in Alaska markets and uses the latest advances in technology to create a greater array of products for urban and rural areas. In fact, in fewer than 10 years, the company has invested $750 million in integrated communication assets. By 2004, GCI will provide high-speed Internet access in all the Alaska communities it serves. Access will be delivered via cable modem, DSL and wireless technologies. Service to rural communities will use state-of-the-art satellite-delivered asynchronous protocols. This $15 million project will give Alaska the rights to proclaim itself as one of the most wired states in the United States. It will serve as a model that other states can emulate to bridge the digital divide between urban and rural communities.

GCI's presence throughout Alaska allows it to give back to the communities it serves. Each year, hundreds of organizations benefit from GCI contributions. Awards are given for proposals that encourage individual growth and positive decision-making. Of special interest to GCI are opportunities afforded Alaska's youth. Emphasis is given to programs where GCI employees have chosen to invest their time and energies.

GCI is proud of the fact that it is always looking ahead to provide innovative services and products Alaskans demand. As a company, it knows its future in Alaska rests on the combination of motivated employees and unsurpassed customer service.

Ted Stevens Anchorage International Airport

Lake Hood and Lake Spenard were home base for floatplanes earlier in the 20th century and now considered the largest floatplane base in the world.

The Ted Stevens Anchorage International Airport is Alaska's principal connection to world commerce as well as the air transportation center of this vast land. It is the leading cargo airport in the Western Hemisphere and the employment center for over 9,000 citizens.

Alaska was not yet a state in 1948 when the 80th Congress appropriated $13 million to fund the development of two Alaskan "international type" airports in Anchorage and Fairbanks. Clearing of the site began immediately and construction was soon underway on two runways. In October of 1953, Anchorage International Airport was dedicated with an 8,400-foot east/west runway, a north/south runway of 5,000 feet, and a modest terminal and tower.

With the dawning of the jet age, Anchorage became the "Air Crossroads of the World."

By 1960, seven international air carriers were using the Airport as a regular stopover on routes between Europe and Japan and between the "Lower 48" and Japan. At the same time, Anchorage was emerging as the focal point for business and economic activity, including interest in Alaska's oil and gas potential. The Airport continued its historic intrastate role of serving diversified transportation requirements of Alaskans for mail, milk, and mineral resources.

In response to international passenger traffic in the 1960s, the Airport constructed an international "hex" terminal at the end of what is now the B Concourse, complete with a busy duty-free concession.

By 1972, the hex was connected to the main terminal and a new ticket lobby and baggage-claim area were added.

This facility now serves as the domestic South Terminal. The following year, a new parallel 10,897-foot east/west runway was completed. A decade later, a 10,496-foot north/south runway was added to the airfield complex and work began on an international facility, now known as the North Terminal, which was completed in time to handle the explosion of international passengers in the 1980s. The 1980s ended with the construction of a 1,200-space parking garage providing more short-term parking space to domestic passengers using the South Terminal.

Meanwhile, globalization of trade provided a stimulus for Pacific Rim time-sensitive as well as general air-cargo shipments. As international businesses adjusted their business models to accommodate new market opportunities, the

International express cargo carriers FedEx and United Parcel Service sort packages and clear U. S. Customs for the burgeoning Pacific Rim trade.

Airport was a major focus of investments by FedEx and United Parcel Service in the express package revolution. The leading express carriers gave international connectivity a jumpstart by establishing major transpacific sorting and customs clearance facilities at the Airport. Today, over 1,500 employees are engaged in these value-added services here.

On July 8, 2001, the Airport was renamed in honor of Alaska's senior United States Senator, Ted Stevens, himself a veteran pilot from the China-Burma-India theater in World War II. It is now the "Ted Stevens Anchorage International Airport," commonly referred to as "ANC."

Alaska's proximity to Europe, Asia, and the Americas continues to make the Airport an efficient stopover for "heavy cargo" international air routes. Cargo aircraft actually increase payload by making refueling stops at ANC on long flights between Asia, North America and Europe. Over 65 percent of ANC's cargo carriers now enhance operational efficiencies through tail-to-tail transfers, freight break down/build up, to full cargo sorting. For years ANC has been the top-ranked air cargo airport in North America for all-cargo aircraft landed weight.

With three full-length, instrumented runways, two passenger terminals, four air parks, 45 wide-body cargo handling and refueling ramps, and the world's largest floatplane base, the Ted Stevens Anchorage International Airport is truly a world-class air transportation center.

All sectors of the air cargo industry, domestic and international, have seen growth. Today the state's population living outside of Anchorage continues to depend on air-cargo service not only for exporting seafood and other resources to market but also for their perishables, consumer goods, mail, emergency supplies, and even fuel for electric power generation.

However, it is the 22 international wide-body cargo carriers that fueled the largest gains in air activity—7 percent average annual increase in landings in the last 10 years, to over 550 per week. The Airport works to remain attractive to the aviation industry. Besides its unparalleled location between three continents, ANC offers all-weather airfield facilities at a significantly lower cost compared to other major world airports. With its supply of ready-to-develop land, aviation companies can lease enough property not just for today's requirements, but for potential expansion to keep up with the marketplace. As a result, private investments in hangars, sorting facilities, and other support facilities in the past 10 years total more than $150 million. Further, ANC fuelers pump more jet fuel into cargo aircraft here than in any other U. S. airport.

The Airport's 50th Anniversary coincides with the 2003 National Centennial of Flight, celebrating aviation since the Wright Brothers' first flight. ANC continues to be a partner with the air industry and its customers who rely on safe, efficient air transportation.

The Airport is responsible for one of every ten jobs in Anchorage. With such a huge commercial influence, the community and State will look to the Ted Stevens Anchorage International Airport as a driving force for economic opportunity.

The Airport celebrated its 50th Anniversary in 2003. Emerging from a domestic point, in the 1950s the Airport began its role serving many of the major passenger carriers in the world for "over the pole" and the first commercial transpacific flights.

Largest among Alaska's thriving airport system, ANC is served by over 20 domestic air carriers and provides vital connectivity to areas of Alaska not served by the road system.

Chugach Electric Association

Customer service and reliability—these are part of the mission Chugach Electric Association has set for itself in meeting the needs of its members. Chugach is Alaska's largest electric utility. It is also an innovative utility, providing numerous services to assist its members.

As an example, in 2001, Chugach began offering several new online programs to its members. One innovative and exciting program is an online home energy analysis. Members can track their energy usage and determine ways to save money.

Other innovative services include the ability to request connects and disconnects online and a program to allow members to pay their bills using an electronic check. These are all geared toward service and making it convenient for members to do business with Chugach at any time of the day or night, all from the comfort of their own home or office.

Chugach members may also pick the day of the month they would like their bills to come due using the flexible payment due-date program, and can even view and pay their bills online with a program called NetPay.

Summary billing is another useful member option. The program creates a single-sheet bill that allows members with multiple accounts to make comparisons and a single monthly payment—while still providing the individual account details on accompanying pages.

Construction performance guarantees assure customers that the utility will complete line extensions or service connections by an agreed-upon date or begin paying them for the inconvenience—assuming the delay is caused by something that is under Chugach's control.

Chugach serves a retail customer base of nearly 60,000 members at more than 73,000 metered locations and provides power to about two-thirds of the homes and businesses in Anchorage. In addition, Chugach delivers power to Alaskans throughout the Railbelt region through

Chugach's Beluga Power Plant is the largest generation facility in Alaska. It is located on the west side of Cook Inlet, 40 air miles away.

Chugach has friendly, helpful customer service representatives who assist members in signing up for service and using various payment programs.

Networks

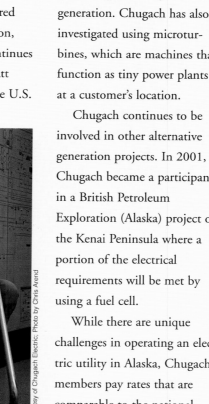

The 138- and 230-kilovolt transmission lines from the Beluga Power Plant are a link to Chugach's members and its wholesale customers throughout Anchorage and the Railbelt.

wholesale and economy energy sales to other utilities from Homer to Fairbanks.

Chugach uses a mix of natural gas and hydroelectric resources to provide clean, reliable, affordable energy. Chugach owns and operates four power plants of its own—including Beluga, the largest power plant in the state. In addition, it is a co-owner of the Eklutna hydroelectric project purchased from the federal government in 1997. Chugach also takes the largest single share of the State-owned Bradley Lake hydroelectric project and dispatches a single-turbine plant owned by another utility. Chugach operates 2,012 miles of transmission and distribution line.

Most of Chugach's power comes from gas-fired turbine generators. In 2001, 86 percent of the kilowatt-hours sold were produced using natural gas as a base fuel for either single- or combined-cycle combustion units. The other 14 percent came from hydro resources.

As most of Chugach's power comes from gas turbines, a steady supply of natural gas is crucial. The utility has long-term contracts with multiple suppliers that will provide a stable, reliable, reasonably priced supply of fuel until about the year 2011.

In addition to its gas-fired and hydroelectric generation, Chugach installed and continues to maintain a one-megawatt commercial fuel cell for the U.S. Postal Service. Chugach is also studying the feasibility of generating electricity from wind generation. Chugach has also investigated using microturbines, which are machines that function as tiny power plants at a customer's location.

Chugach continues to be involved in other alternative generation projects. In 2001, Chugach became a participant in a British Petroleum Exploration (Alaska) project on the Kenai Peninsula where a portion of the electrical requirements will be met by using a fuel cell.

While there are unique challenges in operating an electric utility in Alaska, Chugach members pay rates that are comparable to the national average. In addition, they enjoy reliable service. In 2001,

Chugach averaged 3.66 outage hours per consumer—which translates to power being on 99.96 percent of the time.

Combining innovative customer service along with excellent reliability and reasonable rates allows Chugach to perform the mission it has set out for itself to provide "competitively priced, reliable energy and services through innovation, leadership, and prudent management."

General Manager Joe Griffith leads the Chugach team of 350 full-time, regular employees whose sole purpose is to provide the best electric service possible for members of Alaska's largest electric utility.

Fairbanks International Airport

Fairbanks International Airport features a world-class 11,800-foot (3,598-meter) air carrier runway.

Some 50 years ago, a dirt airstrip kno\wn as Weeks Field served Fairbanks, the transportation and commercial hub of Interior Alaska. As the expanding community outgrew its modest airfield, the Civil Aeronautics Authority (CAA)—forerunner of the Federal Aviation Administration—began building a new airport eight miles south of town.

In 1951, Fairbanks International Airport opened with a paved, 6,000-foot runway as its centerpiece. By 1954, a new terminal building and modern control tower replaced temporary structures. Wien Air Alaska, the airport's primary tenant, constructed a large new hangar. In 1959, the CAA turned the title to Fairbanks International Airport over to the new State of Alaska.

In the ensuing years, Fairbanks International Airport grew to keep pace with a growing community and changing demands. The single largest improvement was lengthening of the runway to 10,300 feet in 1963 to accommodate a growing number of jet aircraft.

The construction of the Trans Alaska Pipeline System in the mid 1970s brought unprecedented traffic to the Interior city's airport. In 1976, boarding passengers totaled 365,000, a record that stood for 19 years. Air freight peaked at more than 200,000 tons in 1974, a number unsurpassed to this day.

Fairbanks International Airport first experienced significant international air cargo traffic in 1978. A west coast jet fuel shortage prompted three Asian cargo carriers to move their freighter operations from Anchorage to Fairbanks. For Japan Air Lines, Korean Air, and China Air Lines, the move meant preserving their Anchorage ration of fuel for rapidly growing passenger services by taking advantage of

Intercontinental air cargo operations at Fairbanks International Airport.

Passenger airlines, boarding over 400,000 travelers per year, link Fairbanks to Canada and the "Lower 48" U.S.

an unlimited fuel supply from the newly opened North Pole Refinery near Fairbanks.

Air cargo operations peaked at over 80 flights per week and, with the passage of the fuel shortage, all three carriers reconsolidated their Alaskan operations in Anchorage. By the mid '80s, intercontinental cargo operations at Fairbanks had dwindled to occasional charters and weather diversions from Anchorage.

In 1986, a public/private partnership, working through the Fairbanks Economic Development Corporation, undertook a market study for Fairbanks International Airport and the next year launched a marketing program on the airport's behalf. This effort, focused primarily on international cargo development, has moved the Fairbanks International Airport from an unranked position as recently as 1990 up to the United States' ninth-largest international air-freight gateway.

An extension of the runway to 11,800 feet, completed in 1997, allows wide-body cargo aircraft to take off year round with unrestricted payloads. Approximately 27 intercontinental cargo flights now pass through Fairbanks each week. The operators include Lufthansa Cargo, Cargolux, Volga-Dnepr, and until recently, Air France. Aircraft service, fueling, and crew changes occur during stopovers on routes between Asia and both Europe and North America.

Growth in domestic passenger service has kept pace with international operations. Now, more than 400,000 passengers annually board aircraft operated by Alaska Airlines and Northwest, as well as more than a half-dozen commuter airlines. In 2001, Condor Thomas Cook Airlines inaugurated non-stop Fairbanks-to-Frankfurt, Germany passenger service—a 'near miracle' of air-service development and a boon to Interior Alaskan tourism interests.

Today, passenger and cargo operations add up to impressive numbers for Fairbanks International Airport, a facility that provides vital transportation links, economic development, and jobs for the Interior Alaska community.

Over 500 small aircraft—on wheels, floats, and skis—are based at Fairbanks International Airport.

FedEx Express

The business world is shifting gears because the New Economy is bringing with it new ways of doing business. The New Economy is high-tech and high-speed; it is fast-cycle, it is networked through e-commerce, and above all, it moves on a global scale.

FedEx Express has consistently led the way in global commerce, providing connectivity to markets that comprise 95 percent of the world's economic activity in just 24 to 48 hours. The FedEx Express facility in Alaska is key to that global connectivity, with flights from Anchorage reaching Tokyo, Frankfurt, and New York in just nine hours or less.

The FedEx Express presence in Alaska has come a long way since opening its hub at Anchorage International Airport in 1989. From 263 monthly flights back then to 465 today and from 139 employees in 1989 to well over 1,300 now, FedEx Express has earned a spot as the ninth-largest private employer in the state. Daily volume for FedEx Express

The opportunity to compete globally requires the resources and the know-how to navigate the vast global marketplace. That is what is done at FedEx Express, and the company does it better than anyone else. Customers have relied on FedEx Express for many years, and the company will remain committed to the people of Alaska.

International Priority—a time-definite, customs-cleared, door-to-door service to more than 200 countries—has grown a phenomenal 300 percent since 1990, the first full year of operation.

FedEx Express is proud of its commitment to the communities in Alaska. The company dedicates significant resources to making the state's communities better places to live and work in, as demonstrated through employee volunteerism and the company's support of local initiatives.

Examples of FedEx Express community involvement in Alaska include:
- March of Dimes
- Pilots for Kids
- Junior Achievement
- Anchorage School Business Partnership
- Anchorage Economic Development Corporation
- Adopt-a-Highway
- United Way

Alaska's Regional Port, Port of Anchorage

Courtesy of Port of Anchorage

Alaska's Corridor of Commerce, Alaska's Regional Port—The Port of Anchorage.

The Port of Anchorage is the northern terminus and critical link in a transportation corridor extending from the Port of Tacoma to the State of Alaska. The Port is the major gateway for Alaska's waterborne commerce. Eighty percent of the state's population receives ninety percent of their consumer goods through the Port of Anchorage. The Port receives more than four million tons of cargo annually, delivered primarily by Horizon Lines and Totem Ocean Trailer Express container vessels. CP Ships/ Lykes Lines has added weekly calls to the Port of Anchorage for its Asia trade lane, Pacific Sprint Service. For the first time, Alaska business can import and export with direct container service through the Port of Anchorage to Korea, Japan, China, and Vancouver, B.C. After passing over the Port's dock's and through its marine terminals, this cargo is then loaded on trucks, trains, barges, and planes and delivered to every corner of the State of Alaska. With connections to every transportation mode serving Alaska, the Port is a hub for a massive multimodal-transportation system that serves the entire state. The Port's valve yards and pipelines are also critical as refined petroleum products are both imported and exported through its facilities. The petroleum needs of Southcentral and Western Alaska are served as are the jet-fuel requirements for Elmendorf Air Force Base and Ted Stevens International Airport.

Tug services are not only available, but are also required since Anchorage has the second-highest tides in the world.

Longshore crews tie up one of the two container vessels that call at the Port two days a week during the winter and three days a week during the summer.

The Port of Anchorage may not be large by international standards, but to the local economy, the impact is huge. The estimated annual economic impact on the state of Alaska by Port operations is $725 million. The Port plays a vital role supplying the raw materials, petroleum/construction equipment, and military supplies that fuel the state's economy. Military planners view the Port as a strategic port and a key component and critical node in all contingency planning. One goal of the Port and its users and customers is to ensure that the movement of goods remains cost effective. This helps Alaskan businesses remain competitive with Lower-48 markets. The safe, efficient and cost-effective operation of the Port will continue to have a positive effect on the quality of life for all Alaskans.

Over the past 40 years, the Port has continued to modernize to keep pace as Alaska's economy has grown and changed. Planners have addressed how the Port of Anchorage must grow to best support the state's economic future, meet the needs of its customers and users, and have the flexibility to accommodate the next generation of vessels that will be making Anchorage Port calls.

Anchorage Mayor Mark Begich has made capital-improvement projects at the Port a priority for his administration. Accordingly, the Port has begun an aggressive, multi-year capital-improvement plan with funding assistance through federal and state grants, Port-retained earnings, and the sale of general-obligation and revenue bonds. The plan includes the construction of a road and rail access to provide for the direct loading of containers on rail cars; a 1,200-foot, multi-purpose dock designed to berth a variety of vessels to include cruise ships; three new 100-foot-gage container cranes; a barge loading facility at the north end of the Port that will incorporate facilities for military load-outs and for heavy equipment, oilfield-module construction, and additional upland development; the reconfiguration of all cargo transit and storage yards; and dredging the harbor to –45 feet. The cost of these projects is estimated to be $227 million, with construction starting in the spring of 2004.

Throughout the Port's history, it has partnered with business, Port customers, and users in the economic changes and diversification impacting lives of residents not only in Anchorage, but also across the entire state. Continued improvement and modernization at the Port is vital as Alaska pursues new opportunities for growth and jobs for future generations.

With new emphasis on the expansion of transportation corridors and the development of the oil, gas, mineral, timber, tourism, and fishing industries, Alaska is poised for future growth throughout the state. This is what "North to the Future" is all about. Opportunities will expand and Alaska will become a better place to live, to work, and to raise a family. Children will want to stay and be a part of what makes the state of Alaska unique.

The Port of Anchorage is a hub of a massive multimodal transportation system that connects over 80 percent of the state by truck, train, and barge and plane connections.

Networks

Avis Rent A Car of Alaska

Courtesy of Port of Anchorage

Courtesy of Avis Rent-a-Car

Courtesy of Avis Rent-a-Car

With all of the challenges of operating in America's largest state, trying harder means a whole lot more for Avis Rent A Car in Alaska. That is because with eight statewide locations, stretching from Fairbanks to Petersburg, trying harder is the only way to ensure Avis's customers are handled with care.

As the only statewide car rental agency in Alaska, Avis is able to provide travelers with seamless services and unmatched consistency in service and rates across an area one-fifth the size of the contiguous United States. The popular "rent it here, leave it there" rentals available from Avis allow customers to customize their own itineraries to maximize their Alaskan adventure.

Avis serves an increasingly large number of tourists drawn to Alaska's vivid scenery and rich culture. These visitors are discovering there is a lot to see on the 14,000 miles of roads in Alaska, from the metropolitan feel of Anchorage to the rustic National Parks and sparkling rivers and lakes across the Last Frontier and from the Summer Solstice in Fairbanks to the breathtaking ports and harbors that provide many a fishing opportunity on the Kenai Peninsula and Southeast Alaska.

Since 1955, Avis Rent A Car has been getting people around Alaska in style.

Regardless of their travel needs, customers have come to expect a high standard of performance from Avis. Service is delivered by a friendly, efficient, and well-trained staff. Despite Alaska's vastness, the company's statewide locations give customers the security of knowing that a friendly Avis office is not far away.

The list of Avis amenities includes: competitive rates for business and leisure; a fleet of over 1,400 late-model vehicles including sport-utility vehicles and vans; Alaska Airlines frequent-flyer miles with qualifying rentals; and a statewide network of locations in Anchorage, Fairbanks, Juneau, Kenai, Kodiak, Petersburg, Sitka, and Skagway.

CARLILE TRANSPORTATION SYSTEMS

Carlile's maintenance crew uses on-board computers that improve safety as well as fuel and maintenance costs.

Carlile Transportation Systems is a national network that leads in industry standards. Carlile's network, along with its highly trained and experienced crew, saves businesses time and money by coordinating virtually every shipping need. Whether the shipment is hazardous waste or an 180,000-pound piece of equipment, Carlile offers fast, affordable pick-up and delivery service.

Carlile also offers options. With four weekly departures from Washington and Alaska, freight is moved from most points within the Continental United States in less than a week. Carlile has daily services within Alaska from Anchorage to Seward, Kenai, Homer, Fairbanks, Valdez, and Prudhoe Bay. Carlile's newest service, Arctic Express, is an expedited Tuesday service to Alaska from Seattle, and Carlile's Alas-Can Express is a seven-day service from Houston, Texas to Edmonton, Alberta, to Alaska.

Equipment is the heart of Carlile's business. With over 1,200 pieces of trucking equipment and 150 tractors, Carlile has some of the best equipment available on the market today. Carlile is also the largest heavy-haul carrier in the state, offering a capacity of up to 125 tons, more than double the capacity of previous lowboys.

If equipment is the heart of Carlile, the crew is its soul. Carlile's customer service philosophy is based on a desire to provide each customer with the best transportation service, whether meeting early delivery requirements or designing special equipment to transport unique freight.

Carlile is active in the United Way, the Alaska Food Bank, the Alaska Trucking Association, and various youth activities and sports teams, among others.

Carlile's future is bright. With the addition of K & W Trucking and Alaska Native-run Kuukpik to the team—and with the commitment to fast, quality service at competitive prices—Carlile will continue to serve not only Alaskans, but also the rest of the country.

Alaskan-owned Carlile Transportation Systems offers cost-effective pricing options, on-site technical assistance, and customized reports for all its customers. With an integrated, multi-modal service, Carlile can expedite shipments to the most remote sites—or just down the street. A well-trained team of logistics professionals provides customers with the best possible price for shipping services via barge, steamship, rail, air, or truck.

Founded in 1980 by Seward residents John and Harry McDonald, the company was started with just two tractors. Over two decades later, Carlile grew to include terminals in eight locations with customers throughout Alaska, Canada, and the Lower 48, and has initiated state-of-the-art web-based tracking.

Each Carlile driver works to ensure that the customer's shipment reaches its destination safely and on time.

Huna Totem Corporation

Tlingit Indian in headgear and regalia.

Huna Totem Corporation (HTC) is an Alaska Native Corporation for the village of Hoonah, located in Southeast Alaska. Formed in 1973 pursuant to the Alaska Natives Claims Settlement Act (ANCSA) and headquartered in Juneau, Alaska, HTC's mission is to provide maximum, assured, equal, and continuing returns to its shareholders, while maintaining control of its lands and recognizing the cultural values of its shareholders.

HTC strives to provide returns to over 1,200 shareholders in the form of corporate dividends. Dividends are paid annually from the company's earnings from a variety of managed investments and operations.

As with many Native Corporations, corporate returns are distributed to a broad shareholder population. Furthermore, none of HTC's shareholders owns equal to or greater than one-half of one percent of the outstanding shares. Therefore, rather than benefiting only a few major owners and key corporate individuals, as is the case with many non-ANCSA companies, HTC benefits numerous shareholders, many of whom live in economically disadvantaged rural areas, and for that reason rely on dividends as part of their livelihood. HTC takes pride in its role of dividend distributor.

As for its investments and operations, HTC currently manages an investment portfolio comprised of stocks and bonds, owns various income-producing properties as well as interest in both venture-capital companies and well-established businesses. In addition, HTC has a small tourism operation that provides cultural interpretation in Glacier Bay on board cruise ships.

Looking forward, HTC will be expanding on its role in the tourism industry through the development of a cruise ship port-of-call facility outside of Hoonah in Icy Strait, which is slated to begin operations in 2004. The facility will offer many shore excursion activities ranging from a Tlingit salmon bake experience to flight seeing, fishing, and whale watching. Tlingit culture is flourishing in the community of Hoonah and will also be featured at the site with artful storytelling and the magic of Tlingit dancing. In addition to augmenting shareholder dividends, this project also creates much-needed jobs.

The company contributes many thousands of dollars each year to the Huna Heritage Foundation, which in turn provides educational assistance to shareholders and their descendants seeking college education, cultural education, and vocational training. The Foundation also promotes and preserves the culture through history recording and stirs the culture in the hearts of the community of Hoonah by hosting Clan Workshops, which immerse the participants in the Tlingit culture.

Totem pole at the Juneau City Museum.

ALASKA AEROSPACE DEVELOPMENT CORPORATION

Aerial view of Kodiak Launch Complex.

At Narrow Cape, on the south side of Kodiak Island, a new high-technology industry for Alaska is taking shape. There are multi-story towers built to shelter rockets as they are readied for launching as well as rocket and payload assembly buildings and launch control centers.

The site bristles with high-tech communications and tracking gear, safety measures that insure rockets stay on course after they are launched. Research and test missiles are launched high from Kodiak over the North Pacific Ocean. Satellites are sent into polar orbit, too.

Narrow Cape, Alaska is the U.S. space industry's "the other cape". It has all the essentials offered by the nation's premier space center, Florida's Cape Kennedy, but it is the only rocket and satellite launch facility not owned by the federal government. That is a key advantage for the U.S. space industry.

Alaska Aerospace Development Corporation, a state corporation, built the Kodiak Launch Complex to offer the commercial space industry, and even the federal government itself, lower-cost options and more flexibility for launches. Kodiak's other big plus is its geographic location at a high latitude on the globe, which means satellites can be placed into polar orbits more efficiently. Because launches are made over thousands of square miles of open ocean south of Kodiak there are no land masses with populated communities, which add up to fewer safety risks.

Although the Kodiak Launch Complex was designed mainly for the commercial space industry, it is also playing an important role in development of the National Missile Defense system. Test missiles launched from Kodiak over the Pacific are being used to test defense radar systems because they follow a trajectory similar to that of an enemy missile attack on the continental U.S. Kodiak will also play a part in tests of missile defense interceptors.

The long-run purpose of the Kodiak Launch Complex is to create a new Alaska industry in space services and support work. Alaska's economy has long been dependent on its natural resources, such as oil and gas, fisheries, and minerals. As Alaska Aerospace Development Corporation further develops the Kodiak Launch Complex, a new, high-tech industry will be created for Alaska.

U.S. Air Force AIT-1 launch, Air Force, November 5, 1998.

Horizon Lines

Anchorage is Horizon Lines' largest Alaska port.

Born of Sea-Land's and CSX Lines' proud heritage, Horizon Lines, the largest and most experienced provider of domestic ocean transportation and logistics services, continues its tradition of superior service to The Last Frontier.

Now a part of the Carlyle Group portfolio, the private investment firm with more than $14 billion under management, Horizon Lines maintains the same industry-leading innovation and high-quality service that made Sea-Land and CSX Lines favorites for Alaskan shippers.

Horizon Lines offers a full range of containerized solutions at competitive rates and convenient sailing schedules for shippers moving cargo between the continental United States and Alaska, Hawaii, Guam, and Puerto Rico. With more than 22,000 containers in a variety of sizes, including advanced-technology refrigerated containers (reefers), insulated containers, vehicle carriers, as well as flatbed and open tops, Horizon Lines has the right equipment to handle the needs of its diverse customer base.

Customers expect innovation from the company that invented intermodalism. Horizon Services Group, developers of the industry's most widely accepted and highly acclaimed comprehensive Internet-based cargo booking, tracking, and billing service, is a member of the Horizon Lines family of services. This unique and innovative company consistently delivers updated technologies, expanded services, and increased efficiencies for its user and customer community.

Horizon Lines' Alaska service offerings include twice-weekly service between Tacoma, Washington and Anchorage and Kodiak, Alaska, and weekly service to/from Dutch Harbor. An extensive network of truck and feeder-barge service connects these ports with Prudhoe Bay, Fairbanks, the Kenai Peninsula, the Pribilof Islands, and other seasonal locations.

Horizon Lines links Dutch Harbor with mainland Alaska and the West Coast.

Hubbard Glacier.

Part Three

Natural Resources

Natural Resources

AMERICAN SEAFOODS GROUP

American Seafoods vessels tied to Pier 90 in Seattle, Washington.

As a global leader in the harvesting, processing, and supplying of quality seafood, American Seafoods Group is one of Alaska's most progressive seafood companies. American Seafoods Group is headquartered in Seattle, Washington with sales offices in Japan and Europe. American Seafoods' fleet is comprised of seven catcher-processors that fish in the Bering Sea and three freezer longliners that fish in the Gulf of Alaska, Aleutian Islands, and the Bering Sea. In addition to American Seafoods' vessel operations, the company operates HACCP-approved production facilities in New Bedford, Maine; Demopolis, Alabama; and Greensboro, Alabama that specifically tailor products that are distributed under a variety of American Seafoods and private-label brands. Due to the company's production capabilities, American Seafoods Group is able to provide the highest quality and freshest seafood possible to its customers.

Bernt Bodal, Chairman and Chief Executive Officer of American Seafoods Group.

The company produces a diverse range of fillet, surimi, roe, and block-cut product offerings made from Alaska pollock, longline cod, sea scallops, and U.S. farm-raised catfish. Finished products are sold worldwide through an extensive global-distribution and customer-support network. The company's core values are centered on responsible stewardship of the oceans by managing sustainable fishery resources and supporting communities where it operates.

"Fisheries play a vital role in the fishing communities throughout the world and especially here in Alaska," stated Bernt O. Bodal, Chairman and CEO. "We are a company that is committed to the long-term sustainability of the resource and of the relationships between our company and the communities in which we operate."

COMMITMENT TO ALASKA'S COMMUNITIES AND RESOURCES
American Seafoods Group believes that sustainable seafood harvests are achievable with little or no adverse ecosystem impacts. The U.S. North Pacific region, for example, is widely regarded as one of the most responsibly and conservatively managed fisheries anywhere in the world. After nearly 30 years of commercial fishing activities, none of the 63 species of groundfish in this region are classified as over-fished, according to the National Marine Fisheries Service.

American Seafoods Group also supports progressive research on marine science and resource management. The Pollock Conservation Cooperative Research Center, for instance, was established in 2000 in conjunction with the University of Alaska. The goal of the Research Center is to improve knowledge about the North Pacific Ocean and Bering Sea through research and education. The Research Center is funded by seven private catcher-processor companies, of which American Seafoods Group is a large contributor. The Research Center donates $1.4 million annually to the University of Alaska, the largest privately funded marine research effort in Alaska's history.

Additionally, American Seafoods Group depends on the Alaskan communities that support its vessel operations through shipyard repairs and vessel resupply services. "Supporting communities in the regions we operate helps us to be a good neighbor," says Bodal, "American Seafoods Group is committed to sharing the benefits of sustainable fishery resources with communities in Alaska."

Southern Pride brand catfish fillets.

The crew of the Ocean Rover *works together.*

The at-sea processor Ocean Rover *sails the Bering Sea.*

Frionor brand bell pepper glazed Alaska Pollock.

Adding Value Through Innovation

The foundation of American Seafoods Group is built upon the capabilities of its sophisticated fleet of fishing vessels and its highly trained crewmembers. To guarantee consistent supply to meet the needs of its customers, all American Seafoods Group vessels are fully equipped with state-of-the-art fishing and processing equipment.

Recent legislation referred to as the American Fisheries Act included a component that revolutionized the Alaska pollock fishery. The cooperative harvesting provision allows fishery participants a specific catch quota. With these quotas in place, fishing strategies changed almost overnight, from a frantic race for fish to one of maximizing the utilization and value of each fish caught. Since co-ops were implemented four years ago, total product utilization has improved a remarkable 49 percent.

Coops also give fishermen the time and individual accountability to minimize bycatch and discard of unwanted species of fish. Bycatch in the Alaska pollock fishery is now only 1 percent, one of the cleanest fisheries anywhere in the world, and discards have been almost completely eliminated.

GROWING TOGETHER IN THE TWENTY-FIRST CENTURY
American Seafoods Group was established in January 2000 with a vision of creating, *From the Ocean to the Plate*, an established global sourcing, selling, marketing, and distribution network bringing quality seafood to consumers worldwide. This network was assembled by merging the harvesting and processing operations that American Seafoods originally formed in 1988 with Frionor USA, a company that offered over 47 years of value-added capabilities. Mr. Bodal, together with Centre Partners Management LLC, spearheaded the formation of the partnership among various investors. Under this enhanced organizational structure, the new parent company now includes two Native-Alaskan non-profit community organizations as investors in the company, representing 21 western Alaskan villages: the Coastal Villages Region Fund of Bethel, Alaska and the Central Bering Sea Fisherman's Association of St. Paul Island, Alaska.

"The employees and management of American Seafoods Group are excited about the future of the fishing industry and Alaska in general," says Bodal. "We look forward to sharing new opportunities and continued success with our partners in Alaska."

Lili Ann, *a Pacific longliner, cruises off the Alaskan coast.*

The Northern Eagle *weathers the storm in the Bering Sea.*

Teck Cominco Alaska Incorporated

Red Dog Mine, located at the foothills of the DeLong Mountains, is a self-sufficient facility with mine, mill, powerhouse, and full living accommodations.

In the early 1970s, members of the NANA Native Corporation looked to the horizon for their dreams. Across the northwest Alaskan tundra, at the foot of the DeLong Mountains, they found one, the hope and promise of Red Dog. With 94 million tonnes of proven and pro ably reserves and zinc grades that reach an extraordinarily high concentration of 20 percent, Red Dog is the largest zinc deposit in the world.

The Alaska Natives Claims Act, settled in the 1970s, empowered Native corporations with ultimate control over the lands that sustain their subsistence lifestyle. Ownership of these lands rich in natural resources also promised economic prosperity. NANA embraced this opportunity and pursued development of the Red Dog Mine.

This was a new era. Indigenous people could now direct, control, and benefit from the resources of their region. Community support to develop the mine needed to be strong.

Grass-roots communication to learn and respond to the people's needs resulted in local endorsement of the mine. NANA's operating partner needed to emphasize local hiring, employee development, and care for the environment. A willingness to share in these responsibilities combined with arctic mining expertise led to the selection of Cominco Ltd. (now Teck Cominco) as a partner.

Red Dog grew from vision to reality and went into operation over ten years ago, in November of 1989. In the first year of production, Red Dog produced 306,127 tonnes of zinc concentrate, and today, Red Dog yields just over one million tonnes. A port and connecting road to the mine were developed with support from the Alaska Industrial Development and Export Authority. The state's $160 million investment is being repaid through Teck Cominco's user and export fees over the life of the mine. The port houses the two largest buildings in the state. They store the concentrates that are then transported by barge to ships waiting offshore during the ice-free summer months.

Red Dog promises a mine life of 40 years. Mineral exploration has revealed several additional deposits similarly rich in zinc, adding new promise for the region. While market prices for zinc fluctuate and challenge the economics of the mining industry, the mineral-rich Red Dog mining district stands strong, facing a long and prosperous future.

The promise of local hire is now a reality for the more than 275 Inupiat Eskimos who are NANA shareholders and work at Red Dog. For these employees, this promise has meant new homes; new boats for subsistence hunting; and the hope, security, and pride they find in economic self sufficiency.

Red Dog has proudly achieved 55 percent shareholder hire. Teck Cominco strives to increase that number and to promote current shareholder employees into more technical and managerial positions through a variety of employee development programs. Whether it is Iditarod musher John Baker helping Teck Cominco promote education by inspiring kindergartners to dream, try, and win; Sonny Adams attending college under a full NANA/Teck Cominco scholarship to learn metallurgical engineering; or Sierra Davis finishing a mine operations apprenticeship on the job; the Red Dog Mine opens doors to the future.

For the Inupiat Eskimos there is a saying, "I walk in one world with two spirits." This is especially true for the Inupiat employee who will work at Red Dog for a one-month period, then return home for two weeks to hunt caribou or to catch salmon to dry on racks as Inupiat ancestors have for generations. For NANA and Teck Cominco, this saying means mining activities must be conducted with care to protect the land, water, and animals that share this world. Sound mining practices, extensive monitoring, and an open dialogue with the local villages ensure that the world of the subsistence community successfully coexists with the Red Dog Mine.

The NANA/Teck Cominco partnership is unique. Industries and indigenous peoples from Alaska to Australia look to Red Dog as a model and a success story. The two corporations continue to evolve and redefine the partnership, stretching past traditional bounds to find the compromises and solutions that make Red Dog a success for all—a star shining brightly above the Arctic Circle.

NANA Native Corporation shareholders hold over 56 percent of the jobs at Red Dog and earn an average of $70,000.00 a year.

PGS Onshore

Equipment and supplies arrive for a job on the North Slope.

Alaskans know a lot about oil. But there are a few things that even the smartest Alaskan may not know about it. For example, where to find it? And once it's found, how much is down there? PGS Onshore is in the business of finding out.

PGS Onshore performs seismic analysis for the oil industry. Seismic analysis is a technical process best explained by Larry Watt, PGS Onshore's Alaska area manager. "What we do is lay geophones and cable on the ground and then we use heavy vibrators that vibrate the ground and send energy down to different sources of information," says Watt. "The returns come up through the cables, we record the information, process it, and produce the data of what's below the surface."

While the method of data retrieval seems straightforward, the actual data is not. Complex blurs and lines stretch out on long rolls of paper and make little sense to the untrained eye. Learning how to read the results, says Watt, "is an entire career."

He points to whorls and blotches, squiggles and dashes. To Watt, these indications of "faults, anomalies, and outcrops" under the surface could mean oil. He indicates a wave on the report. "Oil and gas are trying to come to the surface," he explains. He points to a thick black line hemming in the wave on top. " This could be shale or some other formation that's stopping it from coming up." According to Watt, "there could be hundreds of thousands of barrels."

"Of course", says Watt, "the only final way to find out what's there is to drill." PGS onshore is in the business of providing oil companies with the information to determine, "if and where they want to drill," says Watt. "The rest is up to them."

The company maintains its commitment to capturing quality seismic, even in the coldest climates. PGS stays on the cutting edge—even in the most challenging environments—by having the best equipment, people, and ideas in the business. PGS delivers more data faster, and at higher quality, than anyone else. PGS has invested more than $100 million in the onshore seismic industry's most sophisticated recording and support systems to ensure that the company

An aerial view of a PGS Onshore camp on the North Slope.

The crew spreads the cables and geophones out on a grid.

delivers record-setting performances in any of the world's shallow-water, transition, or onshore zones. Petro-Trac™ Technologies, The Petro Trac suite of seismic technologies, offers clients the opportunity to select multi-component Vertical Cables or surface seismic recording systems. Petro Trac 3C and 9C applications for reservoir characterization and monitoring help locate by-passed reserves, assist in development of production acceleration strategies, and help clients realize the full potential of their fields cost-effectively.

For PGS Onshore, rising to logistical challenges while preserving the environment is a specialty. Recently, a situation arose that resulted in a new way of moving through difficult portions of the Alaska terrain. During the company's first season in the foothills of the Brooks Range, about 100 miles south of Prudhoe Bay, the crews made an unwelcome discovery. Vehicles that were used farther north were of no use in the hilly terrain of the foothills. "The snow doesn't build up on the slope," says Watt, "but where we're working, there are rolling hills, steep hills, and deep gullies."

Not the right environment for the super-wide, 68-inch tundra tires that are so effective for maximum traction and minimal environmental impact on flatter land farther north. "We'd get mired," says Watt. "We would spin out in the deep snow." Caterpillars could pull them out, but at a cost to the environment.

The expert technicians at PGS developed a rubber-track system to replace the tires and equipped the vibrators, recorders, vib-tech units, cable geophone carriers, and trailers with the new system. Two major benefits emerged from their efforts. First, as hoped, the tracking is environment-friendly. The rubber tracking, unlike the steel tires of a Caterpillar, leaves no lasting impact on the tundra surface. "Our goal is to go in, shoot the profiles, and leave and have it look as if we were never there," says Watt. Second, the rubber-tracked vehicles can travel across the crew's "spread"—the cables and geophones laid out on a grid—without doing damage to the equipment.

Currently, PGS operates the only rubber-track crew in the world. However, this will not be true for long. According to Watt, "Anybody doing work in the region will have to use the same equipment developed by PGS."

PGS Onshore may be new to Alaska, but the company has firmly established itself and has a long and prosperous future. "We'll be here as long as people are looking for oil," says Watt. In resource-rich Alaska, that will be a long time indeed.

PGS Onshore developed a specialized vehicle with a rubber-track system that travels efficiently without damage to the environment.

Peak Oilfield Service Company

Ice roads leave no lingering effect on the environment since all evidence of them melts away during the warmer months.

Peak Oilfield Service Company's home office is located in Anchorage, Alaska. In addition to housing its executive management staff, the offices are also the base of several departments including equipment support, procurement, safety and environmental, engineering, Human Resources, business development, accounting, and the administrative support staff. Support for field operations is also provided from this office to Peak's Prudhoe Bay, Valdez, Cook Inlet, and Precision Power operations.

Peak maintains a permanent camp facility at Prudhoe Bay, which includes a 200-person camp, office facilities, equipment repair and maintenance shop, welding shop, fabrication shop, materials warehouse, tire repair shop, steam cleaning and wash facility, and fueling facility. It is also home to one of the largest and most diversified equipment fleets in Alaska, providing equipment support to the North Slope exploration/production and service companies. Peak's equipment is of the type normally employed in civil construction, and drilling and production support. Peak is number one in ice-road and ice-pad construction in Alaska's

Peak Oilfield Service Company hauls drilling waste and cuttings, mud, and water.

Natural Resources

arctic and provides a majority of all drilling-rig moves across the North Slope, as well as in remote locations.

Peak provides tank-cleaning services to the Valdez area oil industries. Peak maintains DOT-certified vacuum trucks, roll-off services, high/low pressure washers, pipe cleaners, and other assorted pumps and cleaning equipment.

Peak provides construction and maintenance services to the Cook Inlet petrochemical industry. Its 6.5-acre site is conveniently located within two miles of the Nikiski Dock, which services Cook Inlet's offshore platforms and facilities on the west side of Cook Inlet and is also well situated to service the oil and gas fields located on the east side of Cook Inlet. Peak's fabrication shops offer top-of-the-line fabrication services for its clients.

Precision Power, a wholly owned subsidiary of Peak, constructs ETL Listed modular power plants that conform with UL STD 2200, NFPA STD 37, and NFPA STD 70. Other services offered include industrial batteries, Uninterruptible Power Systems (UPS), DC power equipment, parts, and accessories.

Peak technicians at work in a fabrication facility.

Skillful technicians are an integral part of Peak's offshore maintenance.

Part Three

BUSINESS, FINANCE AND PROFESSIONAL SERVICES

Okmok Caldera is a volcano located on Umnak Island in the Eastern Aleutians.

DELANEY, WILES, HAYES, GERETY, ELLIS & YOUNG, INC.

Standing in the back row, left to right, are Alex Young, Tim Lamb, Howard Lazar, Donald Thomas, and Jeffrey Stark. Seated in front are Stephen Ellis and Donna Meyers.

Alaska is a state in which growth and development are closely tied to the vast wealth and potential of its natural resources. The discovery of oil in the Cook Inlet region, and subsequent statehood in 1959, marked the beginning of Alaska's emergence as a participant in the global marketplace.

Alaska's growth and increasing sophistication also gave rise to a rapidly changing legal environment. During the past four decades, the law firm of Delaney, Wiles, Hayes, Gerety, Ellis & Young, Inc. has met the challenges of serving the legal needs of a growing state to emerge as one of the preeminent firms in Alaska.

As one of Alaska's oldest and most respected law firms, Delaney Wiles has grown and adapted to meet the rapidly changing legal requirements of its clients. This growth and change has not altered the commitment made by the firm's founding partners to provide superior legal services in exchange for reasonable legal fees.

Today, the firm's 15 attorneys and team of paralegals maintain a practice that serves both the business community and government clients while also serving the vision of its founders. Along the way, former principals and associates of the firm have become judges at the state or federal levels. These include two Alaska Supreme Court justices, three Alaska Superior Court jurists, one Alaska District Court judge, and two who have served on the Federal District Court bench. Another former member of the firm served as Alaska Attorney General, the top law-enforcement officer in the state, and other members of the firm served in the Department of Law, before joining the firm.

"We are extremely proud of our legacy of service to the legal profession," says Stephen M. Ellis, the firm's most senior member. "Among other things,

Mining companies are represented by Delaney Wiles' natural resources section in the acquisition and development of properties throughout the state.

Business, Finance and Professional Services

Delaney, Wiles, Hayes, Gerety, Ellis & Young, Inc. is one of the oldest and most respected firms in Alaska.

we have seen more former attorneys of this firm placed as judges on the bench than any other firm in the state," he notes.

Joining Delaney Wiles in 1975, Ellis has seen the firm develop its practice in three areas. The firm's attorneys handle matters in all types of litigation, as well as transactions involving commercial and natural resources issues. While different members of the firm specialize in each of these areas of practice, there is collaboration among the lawyers in complex cases when it is advantageous to the client.

Delaney Wiles has developed a reputation for its ability to handle complex and difficult litigation. The firm has litigated cases involving product-liability questions, aviation issues, personal injury, professional malpractice, admiralty law, environmental incidents, and insurance coverage. The firm's lawyers have also litigated cases involving construction and operation of the TransAlaska Pipeline System and other oil and gas issues. Delaney Wiles has participated, in one form or other, in a significant amount of major litigated conflicts in the state.

The firm has represented the State of Alaska in a variety of legal proceedings, including the defense of various state agencies in class actions and other tort litigation, arguing the state's interest in major road right-of-way cases, and representation of state officials in civil rights matters. Many of the country's largest insurers, workers' compensation carriers, bonding companies, and several Lloyds of London underwriting syndicates are clients of the firm.

The firm also has a commercial section through which services are provided for a diverse list of clients, ranging from sole proprietorships to international corporations. The state's largest ski resort, an Alaska Native corporation, construction and petroleum industry interests, environmental cleanup organizations, major real estate owners and developers, hospitals, physicians and other professionals, as well as a variety of retail and service industry enterprises, are also among the firm's commercial clients. These varied clients are provided legal services from the acquisition and sale of existing businesses to the start-up of new business ventures. The firm also represents its commercial clients before various governmental agencies, boards, and legislative bodies at the state and local levels.

The natural resources section of Delaney Wiles serves major oil and gas and mining industry clients. The state's rich natural

The midnight sun shines over Kotzebue, Alaska. Attorneys in the natural resources section of Delaney Wiles offer clients an insightful understanding of the history and essence of Alaska's unique natural resource laws and their relationship to the environment. They also advise their clients on the latest developments in federal and state environmental laws.

Delaney Wiles' commercial section provides legal services for a diverse list of clients, ranging from sole proprietorships to international corporations.

resource base offers both opportunity and complex legal issues to companies seeking to develop those resources. This section provides clients with legal guidance in major oil and gas exploration, leasing, development, and marketing projects. Mining companies are also represented in the acquisition and development of properties throughout the state. Attorneys in this section offer clients an insightful understanding of the history and essence of Alaska's unique natural resource laws and their relationship to the environment. They also advise their clients on the latest developments in federal and state environmental laws.

To support its staff in servicing clients, Delaney Wiles maintains an extensive in-house legal library and equally extensive local area network litigation support software including multiple databases, CD ROM-based research materials, Internet research access, Lexis and Westlaw computer research services, and company financial systems. The firm's lawyers are also individually accessible via e-mail on the Internet.

Since 1989, Delaney Wiles has been the only Alaska member of the American Law Firm Association (ALFA), an international association of independent law firms. ALFA's basic objective, through its member firms, is to make available litigation, transactional, and other legal services for a reasonable and cost-effective price, through a network of select independent member firms committed to client service.

Membership in ALFA, also known as ALFA International, is based on a peer evaluation of the candidate law firm and is extended to those firms having a broad spectrum of practice. Membership in the organization links Delaney Wiles with more than 100 member firms across the United States and some two dozen international law firms in 17 foreign countries.

William Moseley and Donna Meyers.

When a national or international legal effort is needed, ALFA firms coordinate their services for the benefit of the client. The firms represent many common clients with common problems.

Closer to home, Delaney Wiles' attorneys provide pro bono work to the community through Alaska Legal Services and religious organizations. The firm's attorneys can be found in neighborhood recreation centers offering legal advice on such topics as wills and landlord/tenant issues. Members also serve on Alaska Bar Association committees and speak to professional associations and service clubs.

Members of the firm have enriched their experiences by involvement in the community through memberships on the Municipal Library Advisory Board, community councils, school curriculum committees, coaching youth ski racers and little league baseball teams, officiating at high school swim meets, serving on national swim committees, participating in local and national cross-country and alpine ski events and bicycle races, and serving as officers and directors of the Alyeska Ski Club, the Alaska Ski Education Foundation, Inc., the Aurora Swim Club, and the Arctic Bicycle Club. They have also been pleased to see their children achieve a number of noteworthy accomplishments, such as becoming a valedictorian of a high school graduating class; achieving status as all-conference athletes; becoming a state wrestling champion, a U.S. Olympic skier, and members of high school championship football and swim teams; and being educated as lawyers, environmental designers, and environmental policy makers.

While the members work to attract top-quality lawyers and clients to the firm, there is no interest in growth for growth's sake. "We try to attract clients interested in furthering commerce and development in the State of Alaska," says Alex Young, the next-senior member of the firm. In recruiting attorneys to the firm, he says, "We hire talent with the expectation they will be with us for years to come."

That approach has worked well for Delaney, Wiles, Hayes, Gerety, Ellis & Young, Inc. The Martindale-Hubbell Law Directory gives the firm its highest rating. Mindful of its founders' commitment to excellence and service, the members look to the new millennium with the confidence that comes with being one of the oldest and most well-established law firms in Alaska.

Oil was one factor that marked the beginning of Alaska's participation in the global marketplace. Delaney, Wiles, Hayes, Gerety, Ellis & Young has the experience and expertise necessary to meet the rapidly changing legal requirements of its clients in this dynamic industry.

Seated from left to right are Alex Young, William Moseley, and Stephen Ellis; standing to the left is Jeffrey Stark.

Alaska USA

Alaska USA Federal Credit Union is Alaska's largest consumer financial institution and one of the largest federal credit unions in the United States. It is a not-for-profit, member-owned cooperative that provides convenient and affordable financial services through state-of-the-art delivery systems.

Alaska USA was chartered in 1948 to serve the financial needs of federal and military personnel that came to Alaska following World War II. It responded to the young state's growth, later serving the financial needs of Trans-Alaska pipeline workers and shareholders of Alaska's Regional Native corporations. Over the years, the credit union's popularity grew along with the membership, and it expanded services throughout Alaska and the rest of the United States as members relocated. This expansion made Internet access and other self-service options a priority. Alaska USA has leveraged these technologies to serve members wherever they work or travel, and to give them access to financial services whenever they want them. In addition, Alaska USA, through its subsidiary Alaska Option Services, has been the leader in providing ATM and point-of-sale services in Alaska.

Today, Alaska USA is the primary provider of consumer credit in Alaska, fueling the economy and helping individuals improve their quality of life through access to high-quality, low-cost financial products and services. While Alaska USA's services and their delivery have changed over the years, its commitment to cooperative credit union principles has remained strong. Alaska USA focuses on satisfying the financial needs of members from all walks of life and levels of income, providing them with the opportunity to be financially successful and to improve their standard of living.

Alaska USA is headquartered in Anchorage's midtown financial district.

Subsidiaries Support Service to Members

In addition to offering savings and loans programs, Alaska USA's commitment to serving members also includes helping them attain homeownership and providing investment options and services. To facilitate access to these services, Alaska USA has two subsidiaries, Alaska USA Mortgage Company and Alaska USA Trust Company.

Branches in retail locations around the state bring convenience to members.

Business, Finance and Professional Services

Alaska USA Mortgage Company provides high-quality service by working closely with borrowers and real-estate professionals.

ALASKA USA MORTGAGE COMPANY

Alaska families need quality, affordable housing, and Alaska USA Mortgage Company assists them in finding the right financing to put home ownership within reach. Alaska USA Mortgage provides comprehensive, cost-effective and professional mortgage services for the purchase or refinance of one-to-four-family residential properties.

Alaska USA Mortgage makes available a full line of mortgage loan products to meet the diversified and unique needs of the Alaska market, Conventional, FHA, VA, Alaska Housing Finance, and Jumbo loans are available, with a wide range of additional options for first-time homebuyers and borrowers with special needs.

Alaska USA Mortgage also works closely with real-estate professionals to ensure high-quality, convenient and responsive mortgage-lending services to its clients. Obtaining a mortgage can be one of life's stressful events, but financing through Alaska USA Mortgage is quick and convenient. Highly experienced mortgage originators guide borrowers through the entire process, working hard to identify the best program to meet each applicant's individual needs. Their commitment to quality is a distinct advantage for borrowers, real-estate professionals and secondary-market investors.

Alaska USA Mortgage utilizes the latest innovations to better serve its clients. For example, it was one of the first mortgage lenders in the state to provide 30-minute mortgage loan approval through Fannie Mae's Desktop Underwriter program. Through servicing options with Alaska USA Federal Credit Union, Alaska USA Mortgage is also able to provide its clients with flexible payment terms and online access to account information and transactions around the clock.

This emphasis on service and innovation has made Alaska USA Mortgage one of the fastest-growing mortgage lenders in the state. It has seven office locations throughout Alaska and has also expanded into Washington State. As an Alaska USA Federal Credit Union subsidiary, it shares the credit union's commitment to quality service, professionalism, affordability and convenience.

Homebuyers can easily negotiate the financing process with the help of knowledgeable professionals at Alaska USA Mortgage Company.

Business, Finance and Professional Services

Alaska USA Trust Company offers professional management of assets and portfolios for investors, leaving them more time for other pursuits.

ALASKA USA TRUST COMPANY

Both institutional and individual investors look to Alaska USA Trust Company for professional expertise and management of their investments. Additionally, personal trust services are available to meet the growing demand of many individuals looking to secure the future for themselves or their children.

- **Individual Investment Services:** More and more people are including mutual-fund investing into their financial plans and Alaska USA Trust has services to meet their individual needs. These include managed-investment accounts, self-directed accounts, individual retirement accounts, and custodial accounts. The availability of these products brings a high level of service and convenience to Alaskans with low minimum investment requirements and modest fees.

- **Personal Trust Services:** Alaska USA Trust offers professional management of trust assets which includes serving in various trustee capacities and providing administrative, custodial, and recordkeeping services for a wide variety of trust relationships. Alaska's liberal trust laws make Alaska USA Trust an ideal choice to provide for the personal trust service needs of both residents and nonresidents. Alaska USA's history of service and stability is a tradition that complements trust administration and management services for today and for future generations.

- **Institutional Services:** Alaska USA Trust specializes in investment custody and securities lending services, helping institutional investors protect and enhance their investment revenues. It offers Alaska public unit investors, such as those associated with state, municipal, or regional governments, personal service through an experienced local institution familiar with Alaska.

Alaska USA Trust also provides custody and securities lending services to financial institutions throughout the country. Alaska USA is a respected name in the credit-union community, where its reputation for quality service and expertise in compliance with credit-union regulatory requirements has provided a solid foundation on which Alaska USA Trust has been built.

State-of-the-art systems, online account access and "real-time" transaction-processing capabilities support Alaska USA Trust's ability to perform for its clients. Professional expertise, long-term relationships and a proven track record make Alaska USA Trust an important addition to Alaska's financial marketplace.

Alaska USA Trust Company provides individualized service to meet investor needs.

Business, Finance and Professional Services

Alaska Option

Alaska Option Services Corporation introduced shared automated teller machine (ATM) service to Alaska in 1983. Today, there are over 350 Alaska Option ATMs in the state, and cardholders have access to their funds anywhere they travel through a variety of regional, national, and international networks. Alaska Option is ranked as one of the largest independent regional networks in the U.S., processing over 22 million transactions in 2002.

ATMs are only part of the Alaska Option story. In 1985, Alaska Option worked with Alaska's dominant grocery retailer to develop one of the earliest debit point-of-sale programs in the country, thus allowing cardholders to pay for purchases with their ATM debit cards. Response to this convenience has been overwhelming. More and more merchants and customers are benefiting from the security, convenience and cost savings associated with electronic financial transactions. Alaska Option cards are now accepted at over 1,600 retail merchant locations statewide and 60,000 merchant locations nationwide.

In 1997, Alaska Option partnered with the State of Alaska in a pilot program to distribute government benefit payments to individuals electronically. As a result of this program, Alaskans can now receive many state and federal benefit payments through the convenience of electronic benefits transfer (EBT). Through Alaska Option and its member institutions, the majority of ATMs and most major retail locations accept state-issued EBT cards for electronic access to government cash benefits.

The Network also offers ATM driving services, card production for both credit and debit cards and settlement/reconciliation services to its member financial institutions.

Alaska Option Services Corporation is overseen by a seven-member board of directors elected from its principal shareholders. Certain administrative and operational services are provided under management contract by its majority shareholder, Alaska USA Federal Credit Union. Its administrative offices are located in Anchorage's midtown financial district. All major Alaska financial institutions participate in the network, representing over 97 percent of the debit cards issued in Alaska.

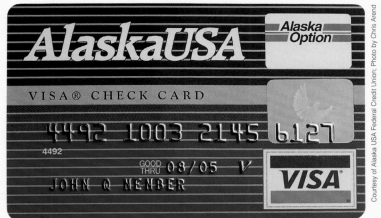

Alaska Option cardholders access funds conveniently at over 350 Alaska Option ATMs statewide and 60,000 merchant locations nationwide.

Above: Alaska Option developed one of the earliest point-of-sale programs in the country.

Premera Blue Cross Blue Shield of Alaska

Premera Blue Cross Blue Shield of Alaska has deep roots in The Great Land, serving the health care coverage needs of businesses, families, and individuals since before Alaska was a state.

Alaskans share a proud heritage. Pioneering spirit and unwavering determination have helped shape The Great Land into a thriving state that still retains its unique character.

As Alaska has grown, Premera Blue Cross Blue Shield of Alaska has been right there, growing too. Its presence in Alaska spans more than 45 years—beginning before statehood. From the very outset, Premera has worked hard to deliver on its mission to provide members greater peace of mind when it comes to health coverage.

Premera Blue Cross Blue Shield of Alaska has continually worked to deliver health-care coverage solutions that preserve the choice and flexibility that Alaskans value. The range of products and services for individuals and businesses encompasses health, dental, pharmacy—even long-term care.

What's more, Premera has fostered strong working relationships with health-care providers, creating a statewide network that helps members and employers save time and money.

Premera's Anchorage office staff understands the unique needs of Alaskans, because they are part of the community.

Business, Finance and Professional Services

Today, Premera Blue Cross Blue Shield of Alaska delivers health-care coverage to over 110,000 Alaskan residents, supported by a dedicated service team exclusively for Alaskan members and employers. For greater strength, Premera draws on the resources of the Premera family of health plans, which serves more than 1.25 million people throughout Alaska and the Pacific Northwest. This regional strength is complemented by local touch: the Anchorage office is staffed by a responsive management, sales, and account services team that calls Alaska home. They understand members' concerns because they are neighbors. Decisions are made promptly, because they are made locally.

Premera's roots in the community also extend to its support of a wide variety of organizations and events that help make Alaska a better, healthier place to live. From the Alaska Health Fair and the local American Cancer Society to the Alaska State Fair and annual Toys for Tots campaign, Premera Blue Cross Blue Shield of Alaska provides support through corporate giving and volunteer service.

The health-care system has changed in Alaska. Today, local physicians and other providers have access to an amazing array of technologies, techniques,

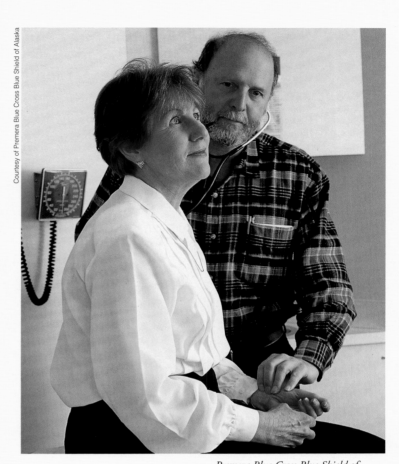

Premera Blue Cross Blue Shield of Alaska provides access to quality care, enhanced by the advantages of a statewide network of private-practice physicians and other health care providers. One example is the high-touch, high-tech practice of Anchorage cardiologist Dr. William Mayer.

Premera's mission is to provide its members greater peace of mind about their health care coverage. It supports this mission by providing superior service and a broad array of products, from health and dental plans to long-term care coverage.

and drugs that enhance the health of Alaskans. These advances also bring challenges, to include keeping access to health care affordable. Premera Blue Cross Blue Shield of Alaska is determined to meet this challenge and works collaboratively with health-care providers, business, and government to develop solutions that work in the unique Alaskan environment.

Premera Blue Cross Blue Shield of Alaska's commitment to meet the challenges of the future is reflected in a new generation of products and services, Premera Dimensions™.

With a wide range of plan designs, freedom of choice, and e-enabled service, Dimensions is ideal for Alaska, which is a state like no other with needs like no other. It is what people expect from a company that listens and is determined to give members the coverage and service they want.

Local touch. Responsive service. Common sense—and compassion. That's what Premera means when they say, "We're here. We're with you."

Even as Premera designs plans and services for a twenty-first century way of doing business, one thing about Premera Blue Cross Blue Shield of Alaska remains stubbornly old-fashioned: its pride in delivering the personal and dedicated service its members expect from a local, caring company.

SEATTLE MORTGAGE (ALASKA)

At Seattle Mortgage (Alaska), "the only thing we'll put you through is your own front door!" Interim construction loans and long-term loans are provided by Seattle Mortgage (Alaska).

Seattle Mortgage (Alaska) is pleased to welcome all new residents and visitors to the wonderful state of Alaska! The company has been in the state since 1986 and has proudly been a part of enhancing Alaska's cities and rural areas by providing convenient, low-cost financing to the state's commercial, construction, and residential customers throughout the years. From assisting first-time homebuyers to relocation clients, it is Seattle Mortgage's goal to set the standard of excellence in this ever-changing market.

The parent company, Seattle Mortgage Company, has been simplifying the mortgage-loan process in more than ten states since 1944. Recently, the company exceeded a portfolio of $2 billion. Seattle Mortgage Company remains one of the strongest truly local institutions in Alaska and the Pacific Northwest. It offers the best loan programs possible in the most expedient time. In fact, Fannie Mae has recognized Seattle Mortgage Company as one of its top affordable-housing lenders!

Because Seattle Mortgage (Alaska) is the only full-line, privately owned mortgage banker in Alaska's marketplace, it can keep up with the latest innovations in technology that provide added convenience. At Seattle Mortgage (Alaska), the friendly and knowledgeable staff is dedicated to the growth of the company and to the success of its clients. It is their mission to do everything possible to merit the client's confidence. Seattle Mortgage (Alaska) looks forward to being of service!

Seattle Mortgage (Alaska) is pleased to offer low interest rates and expertise in local loan origination, processing, underwriting, and closing of one-to-four and multi-unit properties.

Business, Finance and Professional Services

Wells Fargo Bank Alaska

Wells Fargo Bank Alaska's current corporate office at Northern Lights and C Streets in Anchorage.

"We want to satisfy all of our customers' financial needs, help them succeed financially, be the premier provider of financial services in every one of our markets, and be known as one of America's great companies," –Dick Kovacevich, chairman and CEO of Wells Fargo.

Wells Fargo Bank Alaska has established itself as Alaska's leader in the financial services industry and has made an indelible impact on the financial and political history of Alaska. Now more than ever, it is committed to helping fellow Alaskans meet their short-term financial needs and long-term financial goals with friendly, experienced, and knowledgeable bankers and a secure network of financial services.

Wells Fargo Bank Alaska is proud to be involved in the communities where both the corporation and its staff have established roots. Giving more than $1.5 million annually to non-profit organizations across Alaska, Wells Fargo helps fund various programs for educating youth, preserving local art and culture, and providing basic health and human services. Wells Fargo and its employees partner with hundreds of community organizations to revitalize neighborhoods through programs that make housing affordable and stimulate business growth.

From Barrow to Ketchikan, Wells Fargo serves customers in all areas of financial management. Proud to lend its extensive financial expertise to Alaskans far and wide, Wells Fargo has maintained the largest market share in Alaska's financial services industry since the late 1960s. With more than 130 Wells Fargo Automated Teller Machines (ATMs) in Alaska, 51 banking stores, and community agents in three villages, Wells Fargo Bank Alaska reaches more Alaskans than any other financial institution in the state.

The products and services that Wells Fargo offers are cutting edge, earning top marks throughout the industry. *Fortune* magazine lists Wells Fargo & Company as a top 50 "most-admired" company. Most notably, wellsfargo.com is recognized as a top-of-the-line Internet banking site by consumer and business advocate organizations such as Jupiter Research and *Global Finance* magazine. However, as Wells Fargo's company-wide "Service Starts With Me" campaign states, "We strive to put

Wells Fargo contracted with local stage operator Ed Orr, who offered biweekly express and passenger service between Tanana, Fairbanks, and Chitina.

The Anchorage branch of Bank of Alaska at Fourth and E Streets, 1916.

customers at the center of everything we do…our most important product is service."

Wells Fargo & Company is a $334 billion diversified financial services company providing banking, insurance, investments, mortgage, and consumer finance through more than 5,600 stores and other distribution channels across North America as well as internationally.

History

Wells Fargo Bank Alaska maintains the steadfast tradition and character of its roots. It is both a descendent of the first established branch bank in Alaska, Bank of Alaska, and inheritor of the rich heritage of Wells Fargo & Company.

Beginning in 1883, Wells Fargo & Company provided its world-famous banking and express services to Alaskans. From 1911 to 1918, Wells Fargo's daring agents provided reliable express transportation services via stagecoach, steamer, railroad, dogsleds, and horse-drawn sled stages, from as far as California and Mexico and throughout 40 communities in Alaska. All across the Alaskan frontier, Wells Fargo delivered.

"Service is the very backbone of the express," said the company magazine, *Wells Fargo Messenger*, in 1917, "our merchandise is courtesy, willingness, and human ability."

In a wartime measure, the federal government shut down Wells Fargo's operations in Alaska during World War I. Wells Fargo signs disappeared throughout Alaska, not to return until the merger with National Bank of Alaska in 2000.

Meanwhile, Bank of Alaska was beginning its reign as a leader in Alaska banking with the opening of the Skagway branch in 1916. The bank's first president, Andrew Stevenson, envisioned a modern branch-banking system in Alaska.

During the turbulent years of World War I, E. A. Rasmuson took the helm of Bank of Alaska. Over the next eight decades, the Rasmuson family guided the bank as it grew to become Alaska's largest financial institution. E. A., Elmer, and Edward Rasmuson were pioneers of Alaska's financial services industry in personal and business finance management and security. They laid the foundation for the future of banking in Alaska.

In 1920, Bank of Alaska reached $1 million in deposits. Bank of Alaska continued to expand its branch network throughout Alaska and in 1950, a federal charter renamed the bank National Bank of Alaska (NBA).

The Good Friday Earthquake, which devastated Alaska in 1964, destroyed one NBA branch and damaged many others. NBA president Elmer Rasmuson was elected mayor of Anchorage with a long-range revitalization plan for the city.

While under the leadership of Edward Rasmuson, NBA secured the future of banking in Alaska with the establishment of ATM networks and expansion of financial services and branches.

By 1999, National Bank of Alaska remained the state's largest bank with $3 billion in assets and 54 branches. The following year, NBA joined with Wells Fargo & Company, reestablishing the famous Wells Fargo name in the Last Frontier.

Progressive ideas ensure the future of Wells Fargo Bank Alaska in the financial services industry. Wells Fargo is able to provide customers with the vast array of financial services and resources under its belt.

The cover of the MESSENGER, magazine of Wells Fargo & Co.'s Express, 1913. In Alaska, Wells Fargo used dogsleds to carry gold, mail, and express shipments during the winter months.

AeroMap U.S.

LiDAR elevation perspective of Anchorage, June 2002.

Visitors to Alaska are often overwhelmed by contrasts of the Great Land. Most expect to see vast wilderness areas, the grandeur of mountains, glaciers, rivers, and coastal areas. Few are prepared for the apparent time leap from Anchorage, the state's largest city with skyscrapers and all the convenience and bustle of a metropolis, to the austere subsistence lifestyle of Alaska's remote villages. Fewer still really understand the implications of the word "remote" until they comprehend the fact that no connecting roads exist between these scattered communities.

Due to its varied environment and geology, Alaska is an ideal testing ground for technology. The state is home to such extremes as volcanoes and glaciers, barren arctic coasts and immense rainforests—and North America's highest peak happens to be less than 300 miles from the sea. All these geological and environmental extremes are ideal for putting many technologies through their paces, revealing the strengths and weaknesses of equipment and methods. Further, Alaska's extremes require the use of the latest and best technologies to manage cost, time, and quality. This is especially true for mapping.

AeroMap U.S. has thrived in this challenging environment for over 40 years. The company has provided mapping services to military, federal, state, and municipal agencies; engineers; planners; resource managers; the oil and gas industry; miners; and environmental consultants. AeroMap's film library, with nearly 2 million aerial photographs of Alaska's largest cities and smallest villages, chronicles change in the Great Land since before statehood. AeroMap has remained through disasters, natural and man-made, to record transformations that impacted lives of Alaskans.

Making maps from aerial photographs is inherently tied to technology. That was true when the company first began, and it is true now. Today, maps are made with computers for use in computers. AeroMap melds the technologies of cameras and photography with GPS, inertial positioning, lasers, radar, and satellites to find the best mapping solutions for its clients. Further, AeroMap helps many clients build geographic information systems to achieve maximum benefit from their maps. AeroMap uses its unique experiences to provide solutions to geospatial challenges around the world. Like the state of Alaska itself, AeroMap U.S. far exceeds the expectations of visitors.

Business, Finance and Professional Services

Oblique aerial photo of Anchorage, September 1939.

Part Three

BUILDING ALASKA

Craig, Alaska.

ARCTIC SLOPE REGIONAL CORPORATION

One of eight villages in the Arctic Slope Region, Kaktovik is on the north shore of Barter Island, between the Okpilak and Jago Rivers on the Beaufort Sea coast. It lies within the 19.6-million-acre Arctic National Wildlife Refuge. Its approximately 306 residents, mostly Inupiat Eskimos, rely heavily on subsistence hunting and fishing.

Established following passage of the Alaska Native Claims Settlement Act of 1971 (ANCSA), the Arctic Slope Regional Corporation (ASRC) belongs to the Inupiat Eskimo people from the North Slope region of Alaska. ANCSA extinguished claims of aboriginal title to lands in Alaska in exchange for payment of cash and a defined amount of land into a corporate structure. Each Alaskan Native received shares of stock in the corporation that was established to manage the funds and land received from the federal government under ANCSA.

The vision of ASRC is to actively manage its lands, resources, diversified operating subsidiaries, and investments throughout the world in order to enhance Inupiat cultural and economic freedoms. Today, ASRC has over 8,000 shareholders and various investments and operating companies generating annual sales in excess of $1 billion.

With its balance of modern business operations and its support of the traditional values of its Inupiat shareholders, ASRC has developed into one of the true success stories of the ANCSA. The discovery of North Slope oil was a significant contributor to the passage of the Settlement Act because aboriginal land claims had stalled development of that oil. In large part, the Act was passed to allow development to proceed. In some ways, oil affects an older connection to the Inupiat. When the production of hydrocarbons began to replace the demand for whale oil, it lessened pressure on the Bowhead whale, an integral part of the Inupiat culture. While traditional subsistence whaling is now protected and a healthy part of the people's continuing culture, the western commercial whaling that nearly decimated the whale population became obsolete, because oil and plastics made commercial whaling uneconomic and unnecessary.

Natural resource development, especially oil and gas, remains the largest economic force in Alaska and ASRC is a growing player in this industry. The Alaska oil and gas industry is undergoing fundamental changes. Smaller and independent oil companies will play a greater part in the state and ASRC is at the forefront.

Ethel Mekiana makes traditional masks at her home in Anaktuvuk Pass.

RESOURCE DEVELOPER

As part of the ANCSA settlement, ASRC received nearly five million acres of land spread across the North Slope. Most of this land is held in full-fee title with all surface and subsurface rights, while a small part of it is in a "split estate" where the surface rights are held by an ANCSA village corporation. ASRC's land holdings are selected for multiple-use objectives, but with primary resource development potential in mind.

The land holdings contain some of the best quality coal deposits in the world. Located in the northwest part of the state, the Northern Alaska Coal Province shows potential of containing up to four trillion tons of coal. With a significant portion of this deposit on lands owned by ASRC, the coal is clean-burning, high-rank coal with very low sulfur and ash content. These deposits have been explored and test mining has been performed to demonstrate the feasibility of economically producing large quantities of coal in this area. This resource is slated for development and the coal may be shipped to both domestic and foreign markets. In addition, ASRC plans mine-site generation of power for regional economic development.

ASRC also has lands with significant hydrocarbon potential and proven reserves. The Alpine oil field located to the west of Prudhoe Bay near the Colville River Delta is located on lands partially owned by ASRC. Royalties received from production at this field are shared by ASRC with the other eleven Alaska Regional Corporations in accordance with the terms of section 7(i) of ANCSA. In turn, these corporations share a portion of these funds with the village corporations in their region and their at-large shareholders, creating a significant contribution to the state's economy.

Large areas of ASRC's land holdings are being assessed and explored for natural gas reserves in anticipation of the commercialization of other North Slope gas reserves. ASRC is an active proponent of these North Slope commercialization efforts and has performed technical and feasibility work through its various operating business units in support of the on-going gas-study efforts.

Other ASRC lands are being explored for additional oil and natural gas reserves, including holdings in the central Arctic area of the North Slope. ASRC owns over 90,000 acres of subsurface lands in the Coastal Plain of the Arctic National Wildlife Refuge (ANWR). The Kaktovik Village Corporation owns the surface acreage. The ANWR lands are under a long-term exploration agreement with BP and Chevron and are subject to the federal limitations on production and development. ASRC continues to support the opening of this area to oil and gas exploration and development.

Resource development on its own lands, an active role in the overall state oil and gas industry, and other resource activities are all significant parts of the future for ASRC. The company has positioned itself to become the

As part of a drive toward shared services and long-term efficiencies, ASRC and its Anchorage-based subsidiaries relocated to a new building in midtown Anchorage. ASRC's corporate headquarters remain in Barrow.

The blanket toss is a traditional Inupiaq practice for sighting whales off-shore. Although whaling remains an important subsistence activity, the blanket toss is now done mainly for fun at festivals such as Nalukataq, which celebrates the catch of a successful whaling crew.

first truly Alaskan-based and - owned oil and gas company in the state and to meet the needs of other independents seeking local content or a local partner in order to enter the Alaska market.

BUSINESS OWNER

The direct ownership of operating businesses has been the primary focus of ASRC's active business portfolio. While it participates in a number of joint ventures and business affiliations, most of its operations are through wholly owned subsidiaries. These operating activities are grouped into four major categories:

■ Energy Services Group. This Group of companies fulfills operation and support contracts as well as performs construction and engineering services for the major oil-company operators on the North Slope of Alaska. Through this Group, ASRC has maintained more miles of road, built more miles of pipeline, and performed more services than any other company in Alaska. While traditional activities have focused primarily in the Western North Slope oil-producing area on many of the smaller oil-producing units and in support of the transportation infrastructure moving North Slope crude oil to the Port of Valdez, some more recent business development activities have also moved into the Lower 48 states and Canada. The Group operates a major fabrication and instrumentation company on the Gulf Coast, supporting offshore gas development in the Gulf and projects outside the United States. ASRC-designed-and-constructed oil and gas modules and components are operating from the Colville Delta on the North Slope of Alaska, to Sakhalin Island in Russia, to the oil fields of Nigeria. With a major engineering design unit headquartered in Calgary, Alberta, the Group performs engineering design and installation work on a worldwide basis. However, Alaska still remains home and provides the opportunity for a synergistic complement with the company's other Resource Development roles. ASRC's Energy Services Group has become one of the major enablers for smaller independent oil and gas companies seeking entry into the dynamic Alaska market. Rather than competitor, ASRC sees itself as the local partner that can add to the technical and financial expertise that other companies bring to the state. As the mature provinces in Alaska with oil production become more open to independent companies, these opportunities are expected to grow throughout the first decade of the 2000s.

Youngsters and Elders alike wear the traditional fur-trimmed parka as protection from the cold.

Minnie Mekiana takes a break from her summer construction job. Crews take advantage of the long days and relatively warm weather of the short summer season.

ASRC's numerous operating subsidiaries offer a diverse range of services including civil and industrial construction, energy services, and facilities management. Many of these projects also provide job opportunities for ASRC shareholders.

- Petroleum Refining and Marketing Group. This division of ASRC operates two of the major refineries in the state and distributes various fuel and petroleum products across Alaska. A further compliment to the natural-resource focus of the company, it provides in-state value activity for oil produced on the North Slope. With refineries located along the TransAlaska Pipeline System (TAPS), its operations are clean and efficiently operated by retaining the product mix that is appropriate or timely from that crude flow and returning excess crude flow to the TAPS line. Products are sold primarily, at both retail and wholesale levels, into the marine, home heating, and aviation fuel markets, including the large military aircraft operations in the state. It also operates a number of C-Stores that sell and distribute gasoline in state.

- Engineering and Construction Group. An outgrowth of the support for public infrastructure development in the remote villages within its traditional homelands of the Inupiat, ASRC has developed a skilled and successful engineering and construction presence in the Alaska market. Its remote logistics and Arctic experience has made it a leader in the design and construction of public and private facilities in the various communities in Alaska as well as the more remote industrial areas. This Group has constructed schools, health clinics, roads, airports, houses, and various industrial facilities on the North Slope and other parts of the state. Working closely with local communities, these operations have had excellent success in employing the local residents that live in the communities being developed and helping to meet the corporation's vision for economic freedoms for its shareholders by providing employment opportunities. Its engineering operations now have offices in Northwest, Southwest and mid-continent areas of the Lower 48, building upon the experience and expertise developed over many year of operations in Alaska. With a diversified design, planning and management set of services, ASRC is poised to become a major player in the development and expansion of infrastructure in other parts of the country.

- Government Services Group. Beginning as a support contractor to military facilities operating in Alaska, ASRC has become a successful provider of contract and support services to numerous Department of Defense and other agency facilities throughout the world. Often partnering with other experienced government contractors, ASRC has service operations in nearly every state and many foreign countries. It operates one of the largest and most successful conversions of government-operated services to private enterprise for the National Imagery and Mapping Agency in partnership with another Alaska Native Corporation. Taking advantage of the unique status of Alaska Native Corporations, this area of focus continues to grow and become an increasingly significant contributor to the company. Bidding in open and fully competitive government

A crew from SKW/Eskimos, Inc. studies plans for a power plant expansion in Barrow. SKW/Eskimos, Inc., one of ASRC's oldest subsidiaries, partners with many village corporations on local construction projects in the region.

Traditional Inupiaq dancing is very popular and dance groups flourish in each community. Several groups performed in October 2002 at the Open House of ASRC's new offices in Anchorage.

procurements, it has also built an enviable combination of capabilities and contract awards and become one of the country's premier small business success stories. ASRC provides major technical and strategic support to NASA in its development of its future agency role. Working with many major and other smaller specialized service firms, the Government Service operations of ASRC are increasingly sought out by government entities.

PASSIVE INVESTOR

ASRC often invests in larger projects and opportunities that relate to its mission or are complimentary to its other business interests and provide opportunity for significant financial returns. Often focused more on capital appreciation and long-term returns rather than annual operating revenues, these investments have been varied and opportunistic. Building long-term shareholder value can be enhanced with this balance in the portfolio and usually involve taking a small or minority stake in a larger financial venture. ASRC is often able to capitalize on its minority-ownership status, which flows from its Inupiat Eskimo shareholders, in many of these investment opportunities. Where diversity or business development programs encourage minority business participation, ASRC is able to bring its credible business experience and financial capability to participate in a meaningful way with increased likelihood for success. While less visible, this part of the company will continue to support expanded long-term value for its shareholders.

SHAREHOLDER INTERESTS

ASRC serves a fundamental role as the guardian of the rights of its Inupiat Eskimo shareholders in accordance with the intent and purposes of the Alaska Native Claims Settlement Act.

That Act was a compromise and exchange of traditional aboriginal rights for money and land represented by shares of stock in regional and village corporations. Ties to the land remain critically important to the Inupiat people and are a substantial part of this guardian role.

The vision of the corporation reflects the purpose of the Act too, with its reference to "cultural" and "economic" freedoms. The company's focus on jobs and dividends supports the effort to provide its shareholders with freedom to conduct their lives as they desire with the basic needs of life. It was also a significant factor for focusing on development and construction of public facilities in the various traditional communities on the North Slope being funded by public agencies over the last 30 years. This has helped to provide some of the basic services and life-support structure that most of America has

ASRC's corporate headquarters in Barrow.

ASRC Energy Services, the largest of ASRC's operating subsidiaries, offers a diverse range of integrated services for the global energy market.

taken for granted for many years, but simply did not exist for the Inupiat people prior to the land claims settlement.

ASRC has one of the most active and successful shareholder hire programs among all the Alaska Native Corporations and has established an absolute priority that qualified shareholders will be hired first among all its operations. This is a commitment that is seen in the board room of its directors, in its contractual and procurement actions, in agreements with organized labor, and in the actions of the hiring hands in field operations. An emphasis is also placed on training and education for shareholders to ensure that needed skills and qualifications are developed in preparation for assuming positions with the company. An active education foundation ensures all shareholders desiring to go to college or technical school are considered for financial support. ASRC designed, built, and donated a dormitory on the University of Alaska Fairbanks campus to provide a smoother transition of its young shareholders into the university academic environment. An example of a clear vision that directs the company at all levels.

The cultural guardian is also important in supporting other cultural activities and rights of its shareholders. This can be seen in the promotion of subsistence hunting rights for Native Alaskans, its "First People", or in the financial support for cultural and heritage projects on the North Slope and across the state. Traditional cultural values that show respect for elders is demonstrated by the corporation's establishment of a special dividend program for elderly shareholders that did not participate in the traditional cash economy in past years. The tradition of sharing that is evident in the Inupiat's hunting culture, which requires a communal sharing of a successful whale hunt, for example, is demonstrated by the corporation's issuing of shares to children that had been born since the passage of the Alaska Native Claims Act, even though it was a very substantial dilution of the existing shareholders equity in the company.

While only in existence for a single generation, ASRC is a company that expects to be around for many years to come just as its Inupiat shareholders have been as a people for thousands of years.

ASRC OPERATING SUBSIDIARY COMPANIES INCLUDE:
ASCG, Inc.
ASRC Energy Services, Inc.
ASRC Federal Holding Company, LLC.
ASRC Service Center, Inc.
Eskimos, Inc.
Petro Star, Inc.
SKW/Eskimos, Inc.

The warmth and beauty of a summer day in the Arctic – Zack Morry (left), Mark John II (center) and T.J. Ahgook (right) play in a field of wildflowers in Anaktuvuk Pass.

VECO Corporation

The world watched as VECO helped conduct a successful rescue of three whales trapped in the Arctic ice.

VECO Corporation is a world-class engineering, procurement and construction (EPC) company headquartered in Anchorage, Alaska. Providing a full range of EPC services, VECO competes globally with the world's biggest companies in the provision of services to the resource and public sectors.

Organized around a seamless network of regional offices, the VECO team offers clients, "local solutions with world-class expertise," providing services to diverse industries and marketplaces including oil refining, pipelines and terminals, power, oil and gas, federal services and infrastructure, pharmaceutical and biotechnology, and chemical and petrochemicals. VECO is established in the United States, Canada, the Middle and Far East, Africa, South America, and Russia.

Specializing in the challenges of remote and hostile environments, VECO's capabilities range from front-end conceptual engineering studies, process engineering, and feed-package development to multi-discipline engineering; procurement, construction planning, and execution; and operations and maintenance.

VECO considers safety its most important corporate value and strives to maintain an excellent safety record. In 2002, VECO received three Governor Safety Awards of Excellence for engineering and construction.

In response to the need for construction and fabrication services in support of oil and gas development in Alaska, Chairman and CEO Bill Allen founded VECO in 1968. With its beginnings in Cook Inlet, VECO's early history follows the footsteps of the oil and gas industry up to the North Slope. VECO's growth parallels that of the state, which has flourished through the development of natural resources. President Peter Leathard, Chief Financial Officer Roger Chan and key staff have helped shape VECO into one of the largest Alaska-owned companies, employing thousands of people worldwide.

As a pioneer in Alaska's modular construction industry, VECO built the first truckable process modules for the Alaska North Slope oil fields in 1987. Eleven years later, VECO fabricated the largest (5,000-ton) sea-lift modules ever built in Alaska. Following this success, in 2001, VECO constructed

VECO hired over 16,000 workers during a six-month period in 1989 to clean up over 1,000 miles of Alaska shoreline contaminated by oil spilled by the EXXON Valdez tanker.

Building Alaska

The largest modules ever constructed in Alaska were built by VECO in 2001. They are seen leaving the Port of Anchorage on barges on their way to Northstar Island, six miles off the northern shore of Alaska.

ten-story modules that filled three barges with a combined weight of over 13,000 tons.

Fabricated at the Port of Anchorage, the modules were towed to Northstar Island, six miles off the northern shore of Alaska, where they were successfully installed by VECO. The Northstar project was the first off-shore development in Alaska not connected by causeway to the mainland.

In 1988, national and world news media focused on VECO rallying to rescue three whales trapped in the ice near Barrow, Alaska. As the drama unfolded, people everywhere sympathized with the plight of the stranded whales and rooted for their safe release. In cooperation with scientists, the oil industry, and government officials, VECO applied its Arctic expertise, and after two challenging weeks, the whales were finally freed.

In 1989, VECO again captured worldwide attention when it undertook the challenge of cleaning up the largest oil spill in U.S. history. Responding to the catastrophe of the tanker EXXON Valdez, which ran aground in Prince William Sound, VECO oversaw the cleanup of over 1,000 miles of remote and hazardous shoreline impacted by the 250,000-barrel oil spill. As prime contractor, VECO established a complete supply network virtually overnight. Just six months later, at demobilization, VECO had hired over 16,000 people; contracted 2,000 vessels; and purchased and managed billions of dollars of materials, equipment, and labor resources.

Later in 1989, to help protect pro-development interests, VECO ventured into new territory and purchased Anchorage's oldest newspaper, the *Anchorage Times*. Three years later VECO sold the *Times* and negotiated a unique agreement to retain an Op-Ed page. Today, appearing daily in the *Anchorage Daily News*, the "Voice of the Times," speaks to private enterprise, fiscal conservatism, reasonable multi-user development of Alaska's vast resources, and environmentally sound expansion.

VECO strongly supports local community activities in all its regional offices. Generous monetary donations to charitable organizations are complemented by the participation of VECO employees in many altruistic endeavors. Exemplifying VECO's support of non-profit organizations, in 1999, VECO was honored by the City of Anchorage as "Philanthropist of the Year."

In 2001, the governor of Alaska presented VECO the "Exporter of the Year" award in commendation for global export of first-class expertise in engineering, construction and operations (especially pertaining to resource development), cold-weather arctic engineering, and performing in challenging remote locations.

A prior recipient of "Alaskan of the Year" award, Chairman Bill Allen is also ranked prominently with Alaska's most powerful people each year. Bill's youthful dreams of success truly have materialized. VECO today is recognized as one of Alaska's and the world's premier business organizations.

The Women's Run, an annual event that raises money for breast cancer research, is one of the many fundraisers in which VECO employees regularly participate.

H. C. Price Co.

Healy Clean Coal Project.

As a leading participant in some of Alaska's largest construction projects, H. C. Price Co.'s (HCP) presence in the Alaska construction industry has obtained high acclaims. Specializing in the construction of pipelines, process facilities, power plants, utilities, design, and engineering, HCP has proven itself as a top contractor throughout the state.

The company was founded in 1921 when Hal Price borrowed $2,500 to pursue the development of electric-arc welding. Initially, the welding technique was used for tank repairs, but by 1928, HCP had completed a 169-mile pipeline in Texas, thus starting a new era of pipe construction. Establishing a base in Oklahoma, HCP grew to be a major international pipeline constructor. Over the years, Price developed other innovative welding techniques, such as shield-arc welding, removable backup-rings, the "stove Pipe" method of welding, as well as being the first to use pipeline coating systems for buoyancy control.

HCP installed some of the first large-diameter pipelines ever envisioned, thus standing out as the leader of the industry. The 24" Big-Inch Pipeline delivered fuel from the Gulf Coast to Naval Operations on the Atlantic seaboard during World War II, and in the late 1940s, the 30" Biggest-Inch Pipeline extended from the Colorado River to Los Angeles. Such pipelines are commonplace nowadays thanks to Price's pioneering. Since its inception, HCP constructed hundreds of pipelines in North America, the Middle East, North Africa, and Russia.

HCP and its companies have installed more Arctic and Sub-Arctic pipelines than any other pipeline contractor in the Western Hemisphere.

CGF Flare Rebuild: Prudhoe Bay, Alaska.

Trans-Alaskan Pipeline, Section 3.

HCP's first exposure to Arctic conditions came during World War II, when the company built a 1,700-mile oil line in Canada between Fort Norman and Whitehorse to supply fuel to the defenses based in Alaska. HCP then established its Alaska division in 1975, when it was awarded Section 3 of the Trans-Alaskan Pipeline (TAPS), the 144-mile stretch from the Yukon River to Fairbanks. The pipeline crossed two fault zones in this mountainous section, challenging engineers and construction workers alike. After TAPS, HCP continued to perform work in North Slope oil fields, branching out into constructing oil-field facilities and installing modules.

Large Bore Pipe Fabrication, Prudhoe Bay, Alaska.

Throughout its nearly three decades in Alaska, HCP has diversified to become a major constructor of infrastructure projects and electrical-generating utilities. In the late 1980s, HCP successfully bid two projects for the Army Corps of Engineers in Fairbanks. The first project was the expansion of the coal-fired power plant at Eielson Air Force Base, which included a new state-of-the-art instrumentation system. The second project was a major utility expansion project at Fort Wainwright.

HCP went on to construct two major Alaskan power plants from the ground up. The first, a 90-megawatt hydroelectric power plant at Bradley Lake, was remotely located at the head of Kachemak Bay and featured the first and only gas-insulated substation in Alaska. Logistics and barging in a tidal zone area added to the challenge of this project. Next, in 1994, HCP was awarded the construction of the Healy Clean Coal Project, a 50-megawatt power plant near Denali National Park. That project, part of a Department of Energy sponsored program to demonstrate clean-coal-burning technology, introduced new and unique technology, including special coal combustors for maximum burning efficiency and scrubbers to remove sulfur-containing gasses.

In addition to industrial construction, HCP has also been active in other sectors of work. In 1994, FEMA, the government's disaster response agency, commissioned HCP to manage the reconstruction and cleanup of villages flooded by the Koyukuk River. HCP has also had long-term maintenance contracts with Alyeska along the pipeline route and in the Valdez marine terminal.

H. C. Price Co. is proud to promote its diverse abilities and will continue to provide its clients with a reliable source for contracting needs in this time of Alaska's changing economy. While more than a pipeline company, H. C. Price Co. is the only company with such specialized expertise that has maintained continuous construction operations in both Alaska and elsewhere in the world since the construction of the Trans-Alaska Pipeline, and the company expects that record to continue.

ALASKA INTERSTATE CONSTRUCTION, LLC

Outlined by a spectacular view of Cook Inlet, the city of Anchorage, and the Chugach Mountains, this bulldozer is at a site where AIC constructed the dock at Port MacKenzie for the Matanuska-Susitna Borough. The completed dock allows for direct shipments via barge from industries in the Mat-Su Valley.

Alaska's economy depends on natural-resource development to grow and prosper. Today, Alaska is well known for its oil, which supplies over 20 percent of the nation's domestic oil consumption. But Alaska also has large deposits of coal, gold, and other minerals that must be extracted from the earth and refined or processed. The state's resource development companies need reliable, experienced partners to assist them in the process of oilfield and mine development.

Alaska Interstate Construction, LLC (AIC) is among the state's leading partners for resource-development companies doing business in Alaska. AIC has the knowledge and experience to tackle challenging projects in the harshest environments on earth, and excels in every aspect of heavy civil construction for the resource-development industry.

Northstar Island, near Prudhoe Bay, Alaska, is a prime example of AIC's capacity to pioneer a new oilfield and bring it to completion in extreme conditions. The project involved constructing an offshore gravel island over floating ice during the arctic winter months. AIC fabricated special equipment required for the construction effort, including "Snowbirds" to build ice access and haul roads

Northstar Production Island is the only production island in the Arctic. AIC built the gravel island and related infrastructure including slope protection, dock construction, seawall installation, trenching for the sub-sea pipeline, and excavation and placement of foundations and footings for modules. This photo was taken during July at one o'clock in the morning. Once the sea ice is melted, there will be no trace of construction activity.

and extended-reach amphibious backhoes on floats to bury the first sub-sea arctic pipeline.

AIC's construction team excels at finding innovative approaches to solving different development problems. The team has an excellent record of project completion that spans over two decades, and includes many never-been-done-before construction techniques. By becoming involved in a project's development early on and providing engineering support and practical, cost-effective construction solutions, AIC forges a solid partnership with resource development companies.

On Alaska's North Slope, the development of satellite oilfields by major producers is expanding, and by utilizing existing processing facilities, the environmental impact can be minimized. AIC has played a major role in the development of the Tarn, Meltwater, and Palm satellite fields by building roads, drill pads, and bridges to access the new sites. Closer to Anchorage, AIC built the Port MacKenzie dock in Cook Inlet for the Matanuska-Susitna Borough, and continues to construct ongoing expansion projects for the Fort Knox and True North gold mines near Fairbanks and the Red Dog Mine in Northwestern Alaska, the world's largest zinc mine.

Outside the United States, AIC continues to work in the global-resource-development arena in countries as far away as Russia.

In the continuing drive to reduce field-development costs, AIC's experience has demonstrated that the application of a total-project approach can offer significant advantages. The firm's knowledgeable front-end planning group complements the broad-based expertise available within the organization as a whole. AIC's mission is to provide clients with safe, cost-effective, development solutions that meet their business objectives, no matter the size or scope of the project.

AIC's success has been based on the close professional working relationship the firm develops with its clients. AIC has the people and the depth and breadth of experience to successfully execute a broad range of projects, from the small to the major world class, in difficult and challenging environments around the globe.

AIC invented special equipment for the construction of ice roads in the arctic. AIC's "Snowbirds" pump water onto the surface to gradually build ice to enough thickness to support trucks and even an occasional drilling rig. Ice roads are widely used because they melt every spring, leaving no trace of the construction activity. These pumpers are working on sea ice.

A prefabricated home is loaded for shipment from the dock in Cook Inlet at Port MacKenzie. AIC built the dock and related access road for the Matanuska-Susitna Borough.

N. C. Machinery Company

N. C. Machinery Company, the oldest Caterpillar dealer in the Pacific Northwest, has played a vital role in Alaska's growth, including providing machinery and supplies for the Gold Rush in the early 1800s. Its steadfast commitment to providing world-class equipment, parts, technical support, and service to its customers have made N. C. Machinery one of the leading Caterpillar dealers in the world. The Seattle-based company currently serves Washington; Alaska; and Magadan, Russia; as well as parts of Montana, North Dakota, and Wyoming.

The Northern Commercial (N. C.) Company was founded in 1776 by two Russian fur traders, who organized the first company to do business on a continuing basis in Alaska—a fur-trading post on Kodiak Island. The company moved into transportation and mercantile in the early 1800s, and was named Caterpillar dealer for Alaska and the Yukon Territory in 1926. The company's heavy equipment immediately went to work building the basic infrastructure of the new frontier and aiding in the exploration and development of Alaska's vast resources.

In 1994, N. C. Machinery was acquired by third-generation, family-owned Tractor & Equipment Company (T&E). Formerly based in Billings,

Excavators, haul trucks, and other pieces of heavy equipment are used in construction projects throughout Alaska.

One of the more than 100 early trading posts N. C. Co. maintained throughout Alaska

N. C. Machinery supplied over 1000 pieces of Caterpillar equipment for the TransAlaska Pipeline.

Montana, T&E shared N. C.'s history of over 65 years of service as Caterpillar dealers. "The joining of two long lasting industry leaders in the heavy equipment, power generation, and engine business made sense for both companies, as well as our customers," says John Harnish, President and CEO of T&E.

In 2000, the combined company reorganized under the umbrella of Harnish Group, Inc., with N. C. Machinery handling the Caterpillar dealerships and heavy-equipment business in Washington and Alaska, T&E handling the Montana dealerships, N. C. Power Systems handling marine and truck engines and power generators, N C International dealing with expansion into Siberia, and MP&E—the Cat Rental Store—providing smaller machine rentals and industrial supplies.

N. C. Machinery shares rich history with Alaska, and that history exemplifies the company's focus on mutual success with its customers. Caterpillar equipment was instrumental in finding the first Arctic petroleum reserves, as well as building the 800-mile TransAlaska Pipeline. Today, N. C. and its sister companies support all of the primary industries in Alaska, Washington, and Montana with 36 branches, over 1,000 employees, and modern systems to solve support problems overnight.

"During the past century, N. C. Machinery has expanded its services and geographical support in Alaska's developing territories," says Harnish. "Caterpillar equipment has played a major role in supplying villages with electrical power, enabling petroleum engineers to discover oil, supporting miners looking for gold and other natural resources—the list goes on and on."

The opening of N. C. Machinery's Caterpillar dealership in Magadan, Russia now allows the company to sell and service the same Caterpillar equipment on both sides of the Bering Sea. Economies of scale have always provided the company's Alaska customers world-class service and parts-inventory support, and this expansion makes N. C. Machinery even stronger.

The Siberian dealership brings the company full circle, back to its roots in the Russian-American Company formed by two enterprising Siberian fur traders more than two centuries ago.

N. C. Machinery and Caterpillar equipment have played a central role in the development of Alaska's natural resources.

DOWLAND-BACH

During the early 1970s, the Alaska oil boom was in full swing. The oil was there, however, some of the necessary equipment to operate safely was not. Dowland-Bach was founded to meet that need.

The idea started with Ed Clinton, Lynn Johnson and Ron Tharp over Ed's kitchen table. These three men realized that conventional control systems often failed over time in places with extremely cold temperatures and severe winter conditions. Ed, Lynn and Ron felt they had the expertise to build a company to fill that need. They founded Dowland-Bach in 1975. During the last two decades, their vision has turned into one of the few Alaskan-owned and operated companies that specially build, install and maintain control systems for extreme environments.

Ed, Lynn and Ron hired the best technicians, engineers, and electricians. They were people with knowledge of local Alaskan environments. Together, the employees and the founders worked to form a company that can contribute innovative solutions to meet any oil company's needs. Thousands of Dowland-Bach's Wellhead Control Systems have been installed in oil fields — from the freezing icepacks of Prudhoe Bay on Alaska's North Slope, to the damp, humid jungles of South America.

Virtually every Wellhead Control System is specifically designed by Dowland-Bach engineers for each area. The systems use a combination of hydraulic and electrical power to monitor wellhead flow conditions and close the well in the event of an emergency or abnormal operating conditions.

Most of the equipment used for the system is contained in custom stainless-steel enclosures. A major part of the design philosophy is to provide easy maintenance accessibility within the smallest footprint possible. Another factor is proper selection of high-quality, vendor-neutral system components.

Their customers are some of the biggest in the industry, including British Petroleum, Arco and Alyeska Pipeline. Dowland-Bach also provides services for other industries,

Sample of Dowland-Bach Corporation's stainless-steel inventory in the company's Anchorage facility.

Archie Poole of Dowland-Bach Corporation stands next to a custom-manufactured water-injection control panel being built for BP Exploration Colombia Phase II development.

Dowland-Bach's Anchorage-based manufacturing facility includes, among other equipment, a 100-ton x 10' press brake, an environmental test chamber, and a 50-ton iron worker.

such as telecommunications, aviation, construction and local government.

To help achieve the highest-quality service, Dowland-Bach houses one of the largest environmental test chambers in the state of Alaska. The sub-zero freezer is used to double-check manufacturers' specifications, and to collect performance data, before selecting system components. The test chamber can produce temperatures from 350 degrees Fahrenheit to minus 100 degrees Fahrenheit, with cycle times ranging from little more than an hour; or the temperatures can go from one extreme to the other.

"These control systems offer oil production companies an extra measure of protection," says Reed Christensen, Dowland-Bach's general manager. He adds that the Iraqis would not have been successful in setting dozens of oil wells on fire in Kuwait if those wells had been supplied with the company's control systems.

Since the company works almost exclusively with stainless steel, Dowland-Bach was able to expand its original vision to include the distribution of stainless pipe, fittings and metal goods — the same high-quality components and technology the company uses in its own systems. Dowland-Bach is also a distributor of chemical and petroleum processing facility support equipment, and is an ETL and UL listed Original Equipment Manufacturer of industrial control panels, custom enclosures and mini-skids, for use in extreme environments.

Ed, Lynn and Ron struggled for years to build the American Dream. Sadly, too soon after the company had reached that goal, co-founder Ed Clinton passed away in 1997. He is sorely missed by those who knew him and all Dowland-Bach employees strive to make Ed proud of their continuing efforts.

According to Reed, their vision is clear, "We are a company of professional and technical personnel providing design, engineering and manufacturing services with an emphasis on long-term customer satisfaction. We supply and distribute equipment and solutions to a global market."

Those at Dowland-Bach will continue to expand and improve upon this vision in the years to come.

A 150-ton iron worker which is part of Dowland-Bach's Anchorage-based manufacturing facility.

Building Alaska

Superior Plumbing & Heating, Inc.

New heating equipment supplied by Superior Plumbing & Heating was installed for the Arctic Slope Regional Corporation Emerald Building in Anchorage.

No matter how attractive a building is on the exterior, if the interior mechanical system does not work, neither does the building. For 38 years, Alaskans have turned to Superior Plumbing & Heating, Inc. to make sure their projects succeed.

Marion Fox and Bob Pope founded Superior Plumbing & Heating (SP&H) in 1964. In 1978, they hired Jan Van Den Top, now president. "I'm a mechanical engineer by training," Van Den Top said. "They needed a successor and I wanted to put my training into practice. It sounded like the way to do it." Van Den Top found the work challenging. "In a three person operation, you can't specialize in anything. I did a little bit of everything."

Over the years, SP&H has expanded into three divisions—Superior Plumbing & Heating, Alaska Sheet Metal, and the SP&H LINC® Service Division. "We offer the total package," Van Den Top says. "Design, installation and maintenance." With most of the staff participating in an employee stock-ownership plan, everyone pulls together to make projects successful. This diversity and strong commitment to customer service give SP&H what it takes to survive in Alaska's volatile economy.

One of the company's recent accomplishments is Arctic Slope Regional Corporation's Emerald Building—a new ten-story structure in mid-town Anchorage offering 200,000 square feet of Class A office space. Other projects include the Alyeska Prince Hotel in Girdwood, the Alaska SeaLife Center in Seward, a lobby and operating room addition at Alaska Regional Hospital, and a 300-building heating system conversion at Ft. Richardson. Van Den Top found the SeaLife Center project to be a unique experience. SP&H was responsible for piping the tanks, filters and ozone treatment system. "It was an interesting project," he says. "Very unusual."

SP&H's ability to meet the needs of their customers—no matter where the project is located—has placed them consistently among Alaska's top five construction subcontractors. That is a long way from humble beginnings. Success like that can only come from a team of dedicated professionals—professionals who will make sure SP&H continues to get the job done.

Superior Plumbing & Heating installed air conditioning equipment on the roof of the Arctic Slope Regional Corporation Emerald Building.

Part Three

Quality of Life

Johns Hopkins Glacier.

Alaska Native Medical Center

Friends chat in the atrium of the Alaska Native Medical Center. ANMC is a gathering place where Alaska Natives frequently gather to enjoy traditional and contemporary Alaska Native art, learn ways to maintain wellness, or simply sit and visit.

The Alaska Native Medical Center (ANMC) is an Alaska Native-owned-and-managed hospital and a Native place for healing and learning. The Medical Center combines big-city expertise, facilities, and equipment with the individual attention typical of a rural Alaska village. Here, advanced technology joins human caring to provide quality care to the members of 229 Alaska Native tribes.

Located on Tudor Road in Anchorage, ANMC is a 150-bed hospital and a primary-care center jointly managed by the Alaska Native Tribal Health Consortium and Southcentral Foundation. Under the guidance of these two tribal organizations, everything from training to architectural design revolves around providing the best service possible to Alaska Natives, our customer-owners.

The Medical Center has been certified as Alaska's only Level II trauma center, confirming its ability to provide quality care to people with traumatic injuries 24 hours a day, 365 days a year. ANMC has achieved national recognition for its pioneering redesign of primary-care services. This innovative system is centered on patients and their relationship with their care provider. It provides same-day access to care and emphasizes wellness and health promotion.

ANMC is accredited by the Joint Commission on Accreditation of Health Organizations (JCAHO). Awards have come to ANMC from groups such as the Institute of Health Care Improvement, the American College of Physicians, Alaska Emergency Medical Services, and the U.S. Indian Health Service (IHS).

Quality of Life

Nursing intern Peggy Willman talks with patient Luke Maxim. The Consortium offers internships and scholarships for Alaska Natives interested in health careers.

Several physicians have won international or national acclaim for their work in research and treatment of hepatitis, emergency response, and telemedicine. The American Nurses Credentialing Center has awarded Magnet Status to ANMC, one of the highest honors in professional nursing. Many ANMC nurses have already achieved national recognition for their commitment to excellence.

ANMC provides a full range of medical specialties and services with more than 250 board-certified physicians and over 700 nurses on staff.

Parents Barry Lestenkof and Pam James admire daughter Alexandria. The nutrients in traditional foods are especially important to the health of mothers and babies. Consortium researchers look at the health protecting factors that come from breast-feeding and eating subsistence foods.

In 2002, the Medical Center provided more than 360,000 clinic visits, 7,800 admissions, 1,350 newborn deliveries, and nearly 10,000 surgical procedures. ANMC specialists travel year-round to hospitals and health centers in regional hub communities across Alaska to provide on-site clinics. ANMC has extensive residency training and Native leadership development programs.

The Indian Health Service built ANMC in 1997 to replace a downtown Anchorage hospital built in 1953 to serve tuberculosis patients. ANMC now has two major facilities, the main hospital and the Anchorage Native Primary Care Center. These buildings reflect Alaska Native cultures and feature extensive displays of Alaska Native traditional and contemporary art.

The Alaska Native Tribal Health Consortium employs about 1,200 people in the hospital and is responsible for inpatient services, specialty-clinic services, technology and facility support, and clinical support services. These tertiary care services are provided for Alaska Natives referred from all of Alaska.

Southcentral Foundation employs about 700 people at ANMC and is responsible for primary care services for Alaska Native residents of Anchorage and of 55 small villages located throughout Southcentral Alaska.

ANMC boasts an active volunteer Auxiliary, which manages one of Alaska's most highly recognized shops for Alaska Native art and craftwork. ANMC is a Native gathering place where long-time friends can celebrate life events, learn ways to maintain wellness or simply visit one another. The Alaska Native Medical Center is a Native place where people come for healing and learning.

Alaska Native Tribal Health Consortium

The Alaska Native Tribal Health Consortium (ANTHC) is one of Alaska Native Medical Center's two tribal owners. The Consortium was organized in 1997 to run the statewide elements of the Alaska Native health system formerly managed by the federal Indian Health Service (IHS). In 2002, ANTHC employed 1,500 people and had an annual operating budget of $275 million. Principal funding sources for Consortium operations include the IHS, other federal Department of Health and Human Service agencies, and the State of Alaska.

ANTHC is a non-profit organization, owned by the 229 federally recognized tribes in Alaska and Native regional health organizations. The Consortium has a 15-member board of directors, each elected by his or her regional health organization.

The following Alaska Native health agencies are affiliated with the Alaska Native Tribal Health Consortium: Aleutian/Pribilof Islands Association, Arctic Slope Native Association, Bristol Bay Area Health Corporation, Chickaloon Tribe,

Chugachmiut, Copper River Native Association, Eastern Aleutian Tribes, the Native Village of Eklutna, Ketchikan Indian Community, Kenaitze Indian Tribe, Kodiak Area Native Association, Knik Tribe, Maniilaq Association, Metlakatla Indian Community, Mt. Sanford Tribal Consortium, the Ninilchik Village Traditional Council, Norton Sound Health Corporation, Seldovia Village Tribe, Southcentral Foundation, SouthEast Alaska Regional Health Consortium, Tanana Chiefs Conference, the Native Village of Tyonek, Valdez Native Tribe, and the Yukon-Kuskokwim Health Corporation.

Together, these organizations provide a range of medical and community health services for more that 120,000 Alaska Natives. Three-fourths of their customers live in 200 small communities throughout rural Alaska. The Alaska Native health system includes six regional hospitals, ten physician-staffed health centers, and 180 village health clinics. In addition to the Alaska Native Medical Center, the Consortium operates numerous other statewide programs.

Water and sewer facilities and basic health needs are lacking in many Alaska Native villages. ANTHC's Division of Environmental Health and Engineering (DEHE) oversees the expenditure of over $50 million annually for planning, design, and construction of village water and sanitation facilities. DEHE assists local governments in the operation and maintenance of their systems. Engineers deliver these services across Alaska's vast geography and through its challenging climates.

With the Denali Commission (a federal-state partnership providing critical infrastructure in Alaska), ANTHC staff members are constructing and renovating

A rack of fish dries near Kotlik, a village in Southwest Alaska that is accessible only by boat, barge, and airplane—weather permitting. Local waters are frozen nearly nine months of the year, and high winds and poor visibility are common during fall and winter. The Consortium provides health and sanitation services across Alaska's vast terrain and through its challenging climates.

Dr. Mary Totten and Adrianna Stein, Administrative Support, demonstrate how a telemedicine workstation is used to send digital images from rural sites to physicians and specialists in more urban areas. ANMC combines state-of-the-art technology such as computers and satellites with skilled, compassionate care.

small rural health clinics. These facilities are a focal point for health care in hundreds of communities across Alaska. DEHE provides funding and technical assistance for the maintenance and improvement of regional hospitals and health centers. DEHE also conducts environmental-safety inspections for rural health facilities and manages a community-based injury-prevention initiative.

Other important components of public health are research, education, and training. The Consortium's Division of Community Health Services (CHS) Office of Alaska Native Health Research has agreements with IHS and the National Institutes of Health to support research in the areas of cancer, hepatitis, diabetes, and the safety of traditional Native foods. CHS provides technical support and training to affiliated Native health organizations in the areas of dental services, community health aide/practitioner training and certification, substance abuse and mental health, HIV/AIDS services, and immunization services. A new initiative expands training opportunities for village-based dental health aides, behavioral health aides, and personal care attendants to serve rural Native communities statewide.

One of the Consortium's most innovative programs, in its Division of Information/Technology (DIT), works to link healthcare providers in remote villages and clinics with specialists in more urban locations—through telemedicine. Computers, satellites, and custom software allow community health aides to transmit detailed images from an ear exam or EKG, for instance, for expert advice on treatment

Joey Suray, wearing a float-coat for safety, checks out his catch on the Yukon River in Interior Alaska. The Alaska Native Tribal Health Consortium's injury-prevention program educates Alaskans about the importance of using safety gear such as float-coats, seat belts, helmets, and fire alarms.

Workers connect a house to main sewer lines in South Naknek in Western Alaska. Water and sewer facilities, a basic health need, are lacking in many rural Alaska villages. The Consortium designs and constructs water, wastewater, and solid-waste disposal systems, and provides training for the operation and maintenance of village facilities.

Quality of Life

Jennifer Edwards, MD, of the Anchorage Native Primary Care Center, examines Caitlin Stewman in the Family Medicine Clinic. The Anchorage Native Primary Care Center works to establish long-term relationships between the primary care provider and the patient through an open-access program that features same-day access to care.

and diagnosis. ANTHC is managing partner of the Alaska Federal Health Care Access Network (AFHCAN), a comprehensive telemedicine system serving the Department of Defense, the U.S. Coast Guard, the Veterans Administration, and State Public Health Nursing, along with all Native health care facilities. AFHCAN recently won the Grace Hooper Award for technical innovation.

DIT staff also works to ensure that Alaska Native health providers have integrated, state-of-the-art health information systems. Staff provides biomedical engineering services for 70 locations statewide.

The Consortium's Division of Human Resources provides recruiting services for health professionals for all Alaska Native health providers, training and education opportunities on the Alaska Native Health Campus, as well as summer internship and scholarship opportunities for young Alaska Natives pursuing health-related careers.

The vision of the Alaska Native Tribal Health Consortium is to maintain a unified Alaska Native health system by working with Native people and achieving the highest health status in the world—Alaska Natives making healthy choices.

Southcentral Foundation

In 1982, Cook Inlet Region, Inc. (CIRI) formed Southcentral Foundation (SCF), co-owner of the Alaska Native Medical Center, as an Alaska Native non-profit regional health corporation. SCF is part of the CIRI family of non-profit organizations providing educational, healthcare, housing, employment, social, legal, and cultural services to all Alaska Natives and American Indians living in the Cook Inlet region.

In the past twenty years, SCF has built programs distinguished by their innovation, ability to address complex and unique Alaska Native health issues, and responsiveness to its owner-customers. Many SCF programs have received national and regional recognition for innovative, efficient, and effective healthcare services.

More than 45,000 Alaska Natives and their families depend on healthcare and related services SCF provides. Based in Anchorage, SCF employs over 1,000 people in 65 different health- and community-based programs providing services that reach all ages starting in SCF's Early Head Start and Head Start for children ages 0-5 and through primary care at the Alaska Native Medical

Southcentral Foundation provides healthcare and related services to more than 45,000 Alaska Natives and their families.

Kayla Allard, of the Head Start Program, proudly displays her artwork. The goal of the Head Start Program is to help the children and their families achieve their full potential by weaving together culturally relevant education, health, nutrition, and social services.

Center and the Elder Care Program. Other SCF services include optometry and dental clinics, The Pathway Home, Dena A Coy, behavioral health, and community services programs.

Southcentral Foundation serves the members of the Native community in Anchorage and across the Anchorage Service Unit (ASU). This area extends from the Canadian border along the Alaska Range to the west, south to the Pacific and southwest to the end of the Aleutian Chain and the Pribilof Islands in the Bering Sea. Encompassing more than 107,000 miles and including over 50 villages, the ASU presents many logistical challenges in service delivery. SCF supports the local tribal organizations of these areas of the ASU through telemedicine and itinerant physician, dentist, optometrist, pharmacist and other professional services.

SCF recognizes that the treatment of illness is not enough in the journey towards wellness. The ability to get to the underlying issues that truly determine health and wellness requires a different approach. At SCF, the Native emphasis on relationship is understood to be the central value on which to base an effective health system—to fully partner in shared responsibility with the individual, the family, and the community.

This philosophy has led SCF to create a continuum of services that more fully addresses the whole person in context. Programs from before birth, through childhood, adolescence, young adulthood, middle age, old age, and through the death experience are provided. SCF's emphasis on the mental, emotional, and spiritual, in addition to the physical, can be seen in its extensive continuum of behavioral services and its integrated approach. SCF's commitment to children, youth, and elders is apparent in its development of targeted programs for the well and the troubled in these populations.

SCF has also brought together the best of the traditional, the medical, and the "alternative." Its system has tribal doctors, chiropractors, and acupuncturists working alongside mental-health counselors, primary-care providers, medical specialists, and health educators. Traditional ways and Native values are built into SCF's systems design, facilities, educational materials, and interactions with each other. The best of the traditional, the best of the latest technology and science, and the best of system design have come together in a unique and special Native way at SCF.

Much of Southcentral Foundation's success and growth over the past 20 years can be attributed to the hard work and dedication of its employees. SCF continually works to develop and promote its staff, offering in-house training programs and scholarship programs. SCF is 60 percent Native hire and has a fully developed Native-oriented leadership development program.

In its brief 20-year history, SCF has instituted profound changes in its philosophy, which have directly benefited its owner-customers and established SCF as one of the nation's leading healthcare providers for Alaska Natives and American Indians.

Virginia Rowley, LPN, of Southcentral Foundation's Home Based Services Program holds Korrin Winter Nyren in the newly constructed lobby of the Anchorage Native Primary Care Center. Inspired by Native design elements, village life, and Alaska's natural environment, the Primary Care Center contains a large Health Information Center, Native Traditional Healers, Pediatrics, Mental Health, Women's Health, Complementary Medicine, and Family Medicine Clinics.

Providence Health System in Alaska

Providence LifeGuard Alaska has flown thousands of emergency missions in Alaska, Canada, Russia, and the continental U.S.

Just prior to departing for Nome, Alaska, the Sisters of Providence are photographed in Montreal, circa 1902. From left to right: Sister Mary Conrad, Sister Rodrigne, Mother General Mary Antoinette, Sister Napoleon, and Sister Lambert.

100 Years of Caring

The year 2002 held special significance to the Sisters of Providence. It marked their 100-year anniversary in Alaska. To imagine four women traveling to the far reaches of Nome, Alaska in 1902 is remarkable. For the Sisters of Providence, it was their mission—a mission of compassion, caring, and dedication that carried them across a continent to communities in need. Today, Providence honors that partnership with Alaska—a partnership that continues to grow and strengthen as the Sisters of Providence look to meet the needs of Alaskans for generations to come.

A Look Back

Mother Emilie Gamelin led the first group of Providence pioneers from Montreal, Canada via horseback, train, and boat to the city of Nome, Alaska. There, with enthusiastic support from the locals, they established Holy Cross Hospital and began to care for those in need, regardless of their ability to pay.

By 1918, the gold boom was exhausted and most of Nome's 10,000 residents had left the area for more promising destinations. To continue to provide care to those who needed them most, Mother Gamelin and her

Preparing to depart Nome in 1918, the Sisters of Providence traveled via dogsled to Fairbanks.

team left Nome. They traveled to the state's interior to join forces with the Sisters of Providence in Fairbanks who had established St. Joseph Hospital there in 1910, serving 300 patients a year.

In the meantime, construction of the Seward-to-Fairbanks railroad fueled growth and momentum in Anchorage. As this community prospered, its hospital failed to keep pace. By 1935, the *Anchorage Daily Times* reported, "a much larger hospital with more conveniences is sorely needed." In 1939, Providence Hospital opened its doors in Anchorage and became an integral part of what later grew to become Alaska's largest and most vital city.

Here and Now

Today, Providence Alaska Medical Center continues to serve Anchorage-area residents. This 341-bed state-of-the-art medical center is, along with Providence Medical Centers in Seward and Kodiak Island, part of Providence Health System's comprehensive network of experienced healthcare providers and facilities.

World-class specialty-care services include the state's most sophisticated Heart, Cancer, Surgery, Maternity, Rehabilitation, and Behavioral Medicine Centers. Additionally, the Children's Hospital at Providence, a "hospital within a hospital," includes one of the country's newest and most innovative Level III Neonatal Intensive Care Units.

The newly expanded Emergency Services Department at Providence Alaska Medical Center responds to minor and major emergencies 24 hours a day, serving approximately 50,000 patients each year. For critical-care emergencies that occur in remote areas of the state, LifeGuard Alaska provides fixed-wing and helicopter flight service.

Sisters viewing new arrivals at the Providence Maternity Center, 2001.

Quality of Life

The Providence Alaska network of care also includes the Mary Conrad Center, Providence Horizon House, Providence Extended Care Center, Providence Home Health Care, and many collaborative ventures with other area healthcare providers.

LOOKING AHEAD
The Sisters of Providence commitment to providing world-class healthcare, even to those unable to pay for it, stands strong. In 2001, the Sisters of Providence were honored with Alaska's Gold Pan Award for distinguished community service. Last year, nearly 350,000 Alaskans benefited from close to $11 million in charitable services, financial assistance, and community outreach by Providence Health System in Alaska.

The caring mission of the Sisters will always be respectfully embraced by Providence Health System in Alaska. As Alaska and the world continue to change and grow, so too will the good work of the Sisters of Providence —guided always by the light of their mission and core values of compassion, justice, respect for the dignity of persons, excellence, and stewardship.

Providence Alaska Medical Center continues to grow—meeting the needs of Alaskans for generations to come.

THE MISSION OF THE SISTERS OF PROVIDENCE
Providence Health System continues the healing ministry of Jesus in the world of today, with special concern for those who are poor and vulnerable. Working with others in a spirit of loving service, we strive to meet the health needs of people as they journey through life.

St. Joseph Hospital in Fairbanks, circa 1910.

Anchorage School District

As the 86th-largest school district in the country, the Anchorage School District is diverse in its staff, students, and program offerings. Over 50,000 students attend school in the district each day. They are taught by a well-qualified staff that sets high expectations and achieves great results.

Student test scores are better than the national and state averages and graduates earn millions of dollars in scholarships each year. Parents overwhelmingly give teachers and schools high marks. The district is led by a superintendent and school board who are strongly committed to helping all students achieve academic success.

The Anchorage School District student population reflects the rich cultural and ethnic diversity of the community. Anchorage is an educational, medical, and commercial hub for Alaska and is home to many Alaska Natives. Many immigrants from around the globe live in Anchorage and 93 different languages are spoken by district schoolchildren. Roughly 42 percent of students are from ethnic minority groups.

All Anchorage students benefit from a curriculum strong in the core studies of language arts, math, social studies, science, health and physical education. The curriculum is also rich in music and art. Anchorage is a leader among school districts in its optional program offerings. Parents have the opportunity to enroll their children in neighborhood schools with a standard curriculum or in a variety of optional programs focused on science, the arts, or language immersion. Some schools follow specific educational disciplines such as Montessori, open optional, or ABC (back-to-basics). The district is also home to several charter schools.

The Anchorage School District is preparing students to lead the future!

Graduates of Anchorage School District earn millions of dollars in scholarships each year.

Parents have the opportunity to enroll their children in neighborhood schools with a standard curriculum or in a variety of optional programs focused on science, the arts, or language immersion.

ALASKA REGIONAL HOSPITAL

In 2002, Alaska Regional Hospital added a new entrance to the hospital. Designed to be both attractive and functional, this new structure is highlighted by large windows and an indoor fountain to welcome patients and employees with beautiful, natural sights and sounds.

Alaska Regional Hospital was born in 1963 as Anchorage Presbyterian Hospital, located at 8th and L Street downtown. This predecessor to Alaska Regional was a joint venture between local physicians and the Presbyterian Church. In 1976 the hospital moved to its present location on DeBarr Road, and is now a 254-bed licensed and accredited facility.

For 40 years Alaska Regional has been fulfilling the medical needs of Alaskans, continually improving the scope of services and skills available within the state. In 1994, Alaska Regional joined with HCA, giving Alaskans access to the advanced medical resources of one of the nation's largest healthcare providers.

With more than 800 employees and a medical staff of over 400 independent practitioners, Alaska Regional offers a broad spectrum of health services and state-of-the-art technology to the community. The hospital also features full-service heart healthcare, including heart surgery. Alaska Regional is the designated Shriners Clinic facility in Alaska, and free immunization clinics are held monthly for area schoolchildren. Other services include free prostate cancer screenings, health education seminars and support groups for cancer and stroke survivors.

Alaska Regional is the only non-military hospital in the state with its own landing strip, allowing the hospital's LifeFlight Air Ambulance to transport critical care patients from the plane directly the 24-hour Emergency Room. Alaska Regional even offers free valet parking to patients and visitors.

In 1995, 1998 and most recently in 2001, the hospital was accredited by the Joint Commission on Accreditation of Healthcare Organizations (JCAHO), the nation's leading healthcare monitoring organization. Alaska Regional is among the top 4% of hospitals nationwide with its most recent score of 98 out of 100.

Alaska Regional Hospital is dedicated to supporting local events and organizations that promote healthcare and public safety in the state. Some of these include Special Olympics-Team Alaska, American Heart Association, Blood Bank of Alaska, Standing Together Against Rape (STAR), Run for Women, and the 5 Miler for Men's Cancer Research.

Alaska Regional is proud of its ongoing commitment to providing the highest quality healthcare in Alaska. From starting lives to saving lives, one generation to the next. Alaska Regional Hospital.

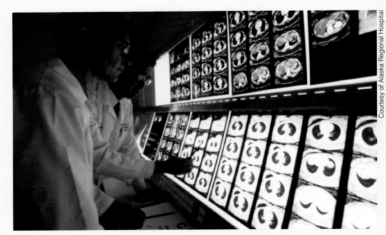

Alaska Regional Hospital is committed to supporting the community's ongoing healthcare needs by investing in the latest technology. In fact, this hospital features equipment and services found nowhere else in the state. Shown here are two physicians reading MRI films.

Alaska Health Resources, LLC

Alaska Health Resources works closely with clients to integrate culturally sensitive, compassionate, patient-centric, and sustainable business practices into healthcare systems.

To build capacity and integrate technology to improve public health, that is the goal of Alaska Health Resources, LLC (AHR). AHR is a consulting group that provides technical assistance and project management services in healthcare planning and improvement practices to individuals, groups, institutions, and governments. AHR works in close collaboration with clients to integrate culturally sensitive, compassionate, patient-centric, and sustainable business practice into healthcare systems. AHR also provides training and education programs for healthcare professionals and administrators. Every project is centered on a commitment to effective, quality, holistic, and affordable health care based on public health standards of primary care and prevention.

The firm offers extensive experience in healthcare planning, administration, and education in hospitals, and academic and corporate settings. Grounded in years of diverse experience, AHR offers practical tools and organizational vision that transforms innovative ideas into effective projects. AHR also operates the Alaska Telecare System, which is used by medical professionals in international arenas and other communities in transition. AHR is implementing state and federal contracts for telehealth protocol and systems evaluation, and consulting with the University of Nebraska College of Nursing for an educational initiative on telehealth and distance delivery.

Telehealth can extend the reach of expert medical care into remote locations and provide a cost-effective network for medical professionals. Development of "virtual clinics" allows communities in transition to access partners, improve skills, and implement health improvement values without the need for substantial investment in infrastructure. It can provide state of the art healthcare systems that are transportable into areas with geographic, transportation, and communication challenges.

REPRESENTATIVE PRODUCTS AND SERVICES INCLUDE:

■ Completing an assessment and feasibility to integrate telehealth services into public and private healthcare facilities.
■ Facilitating networks and technical assistance for international clients to access Alaska's health system, including the export of Alaska's best practices.
■ Providing training programs on business concepts of healthcare and effective primary care.
■ Building local capacity for cost-effective, high quality healthcare services for American and international residents.

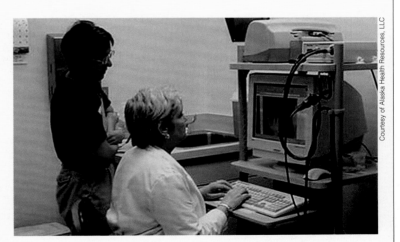

AHR operates the Alaska Telecare System, which is used by medical professionals in international arenas and other communities in transition.

Telehealth extends the reach of expert medical care into remote locations and provides a cost-effective network for medical professionals.

Part Three

MARKETPLACE

Harding Icefield.

Marketplace

Anchorage Chrysler Dodge Center

Mr. Udd recognizes that his success is only as great as the dedicated 140-person staff that runs the large complex on the east side of town.

Anchorage Chrysler Dodge Center circa 1963.

Marketplace

The headline of a framed newspaper article pretty much tells the story of Rod Udd, President of Anchorage Chrysler Dodge Center. "Man known for giving gets award." Still, this tribute is one of countless others sprinkled along with children's artwork on the office wall. There is the National Leadership Award and the Anchorage Downtown Partnership's Pinnacle Award. There is the plaque for being Alaska Republican of the Year and a signed photo with Iditarod musher and champion Jeff King. Draped over a chair is a signed Anchorage Aces hockey jersey (the team is now known as the Alaska Aces due to a recent name change). And on a wall of its own, hangs a picture of him flanked by the President and First Lady. "Last time I saw him he said, 'Hey Rod. How ya' doin'?,'" Udd says proudly. "Of course, I had a name tag on . . ."

Although the shelves and walls of his second-floor corner office barely have space for one more trophy or award or honorary recognition, there are undoubtedly many more in his future that he will always find a place for, as he has over the past 20 years. Small tokens of appreciation for contributions to the community that total in the millions, he would gladly continue to support these events and organizations even if he was the only one who knew of his generosity. To him, it is a two-way street—the community benefits from these sponsorships and it is one way he can thank the community for supporting his business.

"I believe in keeping the money in the community. As long as the community continues to buy the product, I will continue to give back to the community." But, he points out, the art of making money has, by and large, been only the result of his philanthropic beliefs and passion for where he lives and the generations of people he has met in business and through volunteerism.

"Money is not a goal," Udd says. Money is a human life value in a negotiable form. But it comes in handy. If you have a little bit of money, you can help someone; if you don't, you can't."

Just as Udd has come to be regarded as a pillar of the community, so has his business become known as an empire in the auto industry. Anchorage Chrysler Dodge was first located on Fifth Avenue, just west of today's location, where the company operated until 1972 when the Chrysler Building was completed and Chrysler Plymouth moved to the new building and

With a wall filled with awards and honorary recognitions, it is apparent that Rod Udd is a man that is serious about community involvement. For Mr. Udd, to contribute to the community through sponsorships and service is one way to say thank you to the community for supporting his business through the years.

A 2003 aerial of Anchorage Chrysler Dodge Center, "a quarter-mile of vehicles and value."

Marketplace

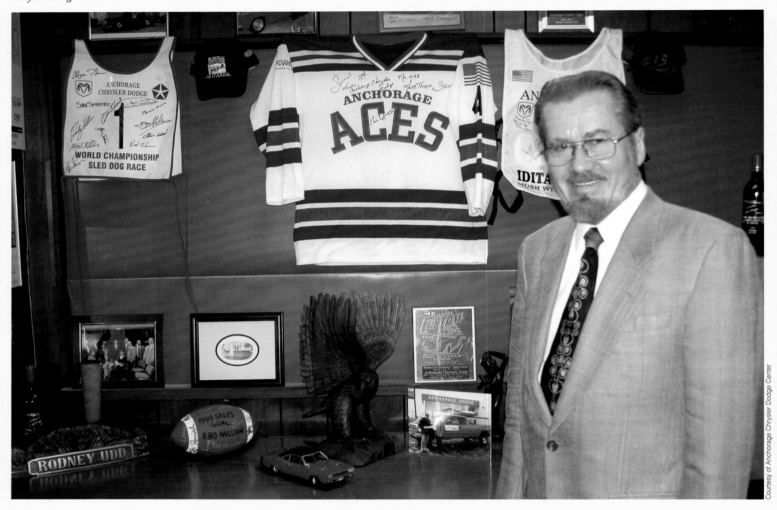

Mr. Udd proudly displays a signed Anchorage Aces jersey in his office. Now called the Alaska Aces, the team enjoys sponsorship from Anchorage Chrysler Dodge.

From left to right, Mrs. Carol Udd, President George W. Bush, Mr. Rod Udd.

Dodge opened right next door. It has since expanded to cover 4.5 contiguous blocks on East Fifth Avenue where Dodge Ram pickups, Chrysler mini vans, and Dodge Durango SUVs line the expansive lot bordering this busy commuter corridor.

Founded in 1963 as Fifth Avenue Chrysler by Kenneth B. Davis and Glen Phillips, Phillips sold his share, making Davis the sole owner in 1965. At the time, Anchorage had over 90,000 residents and was well on its way to becoming the headquarters for the Alaska oil industry. Eighteen oil companies had already opened offices in Anchorage and Anchorage Chrysler Dodge was poised to sell vehicles to the thousands of employees and families moving to Alaska to ride the first boom of the oil industry.

The son of missionaries, Udd has lived and traveled all over the world. As a child, he had the opportunity to live in Africa for several years. Although he especially enjoyed the African culture and still talks about the anthropology and natural history of that continent, he says the idea of following in his parents' footsteps did not challenge him. He was more interested in math and the sciences. That, combined with his interest in people and the gift of gab, made him a natural in sales, and he quickly advanced his career, first by selling life insurance in the Seattle area and then by working for several car dealerships, eventually landing at Anchorage Chrysler Dodge in 1973.

From sales, he was promoted to sales manager, then to fleet manager, then to the head of inventory control, before he finally bought the dealership in 1989. Since then, he has grown the business into one of Alaska's largest dealerships, with annual revenues of over $64 million. This success is not based on any secret formulas or cutthroat marketing strategies, he says pragmatically.

First, he says, Chrysler continues to develop reasonably priced, solid products, specific to the needs of Alaskan drivers. " 'Promote quality at a fair price', is the first principle behind Anchorage Chrysler Dodge," Udd says. "And it's worked out well. Ever since I first learned to sell, I always thought it was more important for the customer to know what

Rod Udd and his wife Carol pose with 2003 Iditarod Champion Robert Sorlie.

Mr. Rod Udd presents a Dodge Ram and the Joe Redington Sr. Winner's Trophy to four-time Iditarod Champion Doug Swingley.

" 'Promote quality at a fair price', is the first principle behind Anchorage Chrysler Dodge," say Rod Udd.

Time Magazine's *2001 Quality Dealer Award* was presented to Mr. Rod Udd, owner and president of Anchorage Chrysler Dodge Center.

they were buying in terms of quality. Instead of advertising a huge discount, we tell them about the product. People don't buy discounts, they buy quality at a fair price."

At the same time, Udd recognizes that his success is only as great as the dedicated 140-person staff that runs the large complex on the east side of town. It is this basic product–service balance that continues to keep his company in the lead when it comes to sales numbers among Chrysler dealerships. Another key ingredient that contributes to building and maintaining employee morale and customer satisfaction is being involved in the community's well being.

The success of Anchorage Chrysler Dodge translates into a broad range of support for community events and charitable causes. The most notable of these is the dealership's annual sponsorship of the Iditarod Trail Sled Dog Race. The 1,110-mile Anchorage-to-Nome event draws about 75 dog teams and mushers from Alaska and around the world. Udd became especially interested in helping the race when other sponsors began withdrawing their support during the late 1980s due to mounting political pressure from animal activists. In addition to donating about $300,000 to the annual event, Udd also gives the winner of the race a Dodge Ram pickup.

Always looking out for Anchorage, in 1999 Udd called the local television station to find out why the Anchorage Fur Rondy World Championship Sled Dog Races were no longer broadcast. It was due to lack of sponsorship, he was told. Always happy to help, he assisted in getting the races back on the air and has since agreed to donate a Dodge Durango to the winner. Other beneficiaries of Anchorage Chrysler Dodge support include Big Brothers/Big Sisters; the Boys & Girls Club; the Alaska Aces Hockey Team; Anchorage Downtown Partnership; Habitat for Humanity; Cook Inlet Soroptomists; the Lions Club; and Intervention Help Line, a local drug and alcohol counseling program.

Despite his love of adventure and travel, the things that keep Rod Udd in Anchorage are the small-town flavor and the unlimited opportunities. Says Udd, "I've been to most places on earth. This is a place where you have a lot of challenges, but for anyone who enjoys a challenge, there is a lot of opportunity and a good future for those who don't mind working hard. I find a great deal of satisfaction in contributing to the well being of those around me."

Still, Udd hopes to grow the company into an even larger presence on Fifth Avenue and eventually open other locations in and around Anchorage. It is this futuristic thinking that recently increased the dealership's visibility via the Internet. Alaskan customers who live in rural bush communities can now access from their computers everything from the vehicles on the lot of Anchorage Chrysler Dodge to sales-and-lease options, incentives, information on service and parts, employment opportunities, pre-owned vehicles, and commercial fleet sales.

Looking back, he says, he has sold cars to generations of families. He has started successful careers for generations of employees. And he has raised generations within his own family. Just as important to him, though, are residents he may never have met, but who he has had the opportunity to contribute to by supporting business development, recreation, community events, and healthy families.

Mr. Rod Udd, owner and president of Anchorage Chrysler Dodge Center, awards Anchorage Aces "Player of the Month" trophy to Paul Williams.

Marketplace

HISTORIC ANCHORAGE HOTEL

Guests relax in the ambiance of the Anchorage Hotel's European-style lobby after a day of sightseeing or meetings.

In the heart of Anchorage's bustling downtown sits a unique gem among hotel properties. Anchorage was barely a year old when Hotel Anchorage began operation. Built on the corner of 3rd Avenue and E Street, Hotel Anchorage became the central gathering place for the new city. Over the years, the hotel became well known as the premier place to stay in Anchorage. By 1936, an expansion was needed and a new building, the "Annex," was built across the alley and was connected to the original building by a skybridge.

Many notable people have stayed at Hotel Anchorage over the years. Will Rogers and Wiley Post stayed at the hotel before their journey to Fairbanks and subsequent ill-fated flight. But perhaps the most notable guest was well-known Alaskan landscape artist Sydney Laurence. Mr. Laurence lived in the original hotel as well as the annex for many years. He painted his famous landscape of Mt. McKinley in the lobby of the original building and ultimately exchanged his masterpiece for a year's worth of rent. The painting is now on display at the Anchorage Museum of History and Art.

On Good Friday, March 27, 1964, Southcentral Alaska was struck with the most devastating earthquake ever to hit North America; it registered 9.2 on the Richter Scale. Unlike many of her surrounding original townsite buildings, the Anchorage Hotel survived the four-and-one-half minutes of tumultuous movement with very little damage. Today, the only telltale sign of that traumatic event is a slight shift in the foundation.

In 1969, the original hotel building was sold; however, Hotel Anchorage Annex continued to operate for several years before being purchased in 1988 by its current owner. After an extensive restoration project, the Historic Anchorage Hotel reopened in 1989 as Anchorage's premier boutique property, and in April of 1999, the Historic Anchorage Hotel was officially placed on the National Register of Historic Places.

Located on the starting lines of both the World Championship Dog Sled Race and the world-famous Iditarod Dog Sled Race, the Historic Anchorage Hotel is the ideal location for any visit to Anchorage, in winter as well as summer! Lavish amenities are included with every stay.

Owned and operated by local, lifelong Alaskans, the Historic Anchorage Hotel is one of four businesses under the Grizzly's, Inc. umbrella. The Historic Anchorage Hotel, Rumrunner's Old Towne Bar, Grizzly's Gifts, and Global Image are all-important facets of Anchorage's economy and community support.

The Anchorage Hotel's Queen Ann styling welcomes guests and gives a true sense of Anchorage's history.

Alaska Center for the Performing Arts

Anchorage Opera.

The Alaska Center for the Performing Arts, prominently situated in the middle of downtown Anchorage next to Town Square Park, has become a cultural landmark in Alaska. The Alaska Center has hosted everything from big Broadway shows to important lecturers, from weddings and banquets to ballet and jazz, and from comedians to classical music. This world-class facility's three theatres, spacious adjoining lobbies, and support areas comprise 176,000 square feet, all within a one-block area.

One of Anchorage's most unique gathering places, the Alaska Center is where people come to celebrate the performing arts and other important social occasions. Each year the Alaska Center hosts over 600 performances and events featuring over 4,000 amateur and professional performers and attended by over 250,000 patrons.

Since opening its doors in 1988, the Alaska Center for the Performing Arts has hosted a wide variety of performances featuring both local and internationally acclaimed performers and productions. National touring productions like *Cats*, *Les Miserables*, *A Chorus Line*, *The Phantom Of The Opera*, and *Fame* have graced its stages, and a "who's who" of national and international stars like Itzhak Perlman; Gregory Hines; Yo Yo Ma; Ray Charles; Tony Bennett; Mikhail Baryshnikov; B.B. King; and Crosby, Stills, and Nash have appeared at the Center. Each year, the Center's eight resident companies and several other local performing arts groups treat Anchorage audiences to an outstanding array of Broadway musicals, dance performances, choral music, opera, symphony, popular music, drama, and much more.

A nighttime view of the world-class Alaska Center for the Performing Arts.

ASPEN HOTELS

The Aspen Hotels are Alaska's newest hotel chain. Founded by George Swift and Carol Fraser in 1998, the company became a reality because of a desire on their part to build a first-class hotel chain in Alaska. As a beginning, the two purchased land in Juneau and, in 1999, construction began on their 94-room hotel.

The same year brought the purchase of the Valdez Village Inn—a great, yet older, property in downtown Valdez. Carol and George immediately closed it for massive renovation. In March 2000, the 78-room hotel was reopened, with a marvelous new look inside and out.

Fairbanks, the Golden Heart City, is renown for its 24 hours of summer sun. More hotel rooms were required to accommodate the growing Interior city's busy summer season, so in 2000, a new Aspen Hotel was being built close to the airport. That winter, the fortunate construction crew encountered unseasonably mild temperatures, allowing for an early opening of the 97-room hotel in April 2001.

When people think of salmon, they immediately think of the world-famous Kenai River. Many memories and exaggerated stories begin on this river . . . but what a great location for an Aspen Hotel! In 2001, construction started on Soldotna's first Aspen Hotel on the banks of the Kenai River. It opened in March 2002 with the only hotel swimming pool on the entire Kenai Peninsula and is now home to many Alaskan sports enthusiasts who spend their weekends competing at the Soldotna Sports Center.

In 2001, construction began on a downtown Aspen Hotel that was to be the crowning jewel of the Aspen chain. The four-story hotel would be complete with an elegant lobby, presidential suite, and many amenities not found in its four sister properties. As the hotel was nearing completion, a fire was started in the lobby that destroyed the entire structure.

In 1999, construction began on the first of the Aspen Hotels in Juneau. Today, the cheerful lobby welcomes guests to the comfortable 94-room hotel.

Through such a tragedy, Alaskan's show their true characters. Carol and George received over 400 emails, faxes, and phone calls from people who wanted to help during the crisis. It was truly a humbling and overwhelming time for them. With support from their neighbors, the decision to rebuild was made and construction was restarted two weeks after the fire.

January 23, 2003 was the day the Anchorage Aspen Hotel opened its doors. The Aspen Hotels is a proudly Alaskan-owned and -operated chain, bringing first-class hotel rooms to its Alaskan neighbors.

The city of Fairbanks is well known for its 24 hours of summer sun, and its busy summer tourist season continues to grow because of it. The Aspen Hotel in Fairbanks was built to help accommodate the many visitors to the Golden Heart City.

ANCHORAGE FUR RENDEZVOUS

Anchorage Fur Rendezvous is Alaska's Premier Winter Festival. For seventeen days beginning in February, Anchorage hosts over 120 events including the World Championship Sled Dog Race, Snow Sculpture Competition, Fur Auction, Dog Weight Pull, Grand Parade, Snowshoe Softball, Grand Prix Auto Race, Carnival, and Fireworks Extravaganza. There is truly something for everyone with a sense of adventure and a fun-loving attitude.

The first Festival took place in 1935. Miners and trappers would meet in Anchorage to sell furs and minerals for cash and goods. It was also a time to celebrate the "beginning of the end of winter!!" At that time, the fur auction was the main event, but the emphasis today has broadened to include cultural, educational, and historical activities.

Over the years, Festival participation flourished and events were added to accommodate a growing population. In 1955, Greater Anchorage, Inc., a not-for-profit organization, was formed to produce the annual Festival. Its goal was to contribute to quality of life by providing entertainment, education, and cultural growth for Alaskans through fun and exciting mid-winter events.

Since the beginning of the Festival, Anchorage has seen significant changes. Tourism has become a significant backbone in Alaska's economy. Fur "Rondy" is a rare wintertime event that brings a much-needed boost to the long days of winter and to an otherwise quiet visitor season. Thousands of visitors from around the state and Lower 48 make their way to Anchorage to join in the celebration.

Today, the purpose of the Festival is to create fun and vibrant events for Alaska's residents and visitors so they enjoy themselves. Winters are a challenge to some and "Rondy" is a vital and welcome annual tradition to the community and its economy. "Let's Rondy!!"

Mr. Rod Udd congratulates the winner of Anchorage Fur Rendezvous' World Championship Dog Sled Race. The jubilant winner received a brand new red pickup courtesy of Anchorage Chrysler Dodge Center.

Fur Rendezvous' Miners & Trappers Ball.

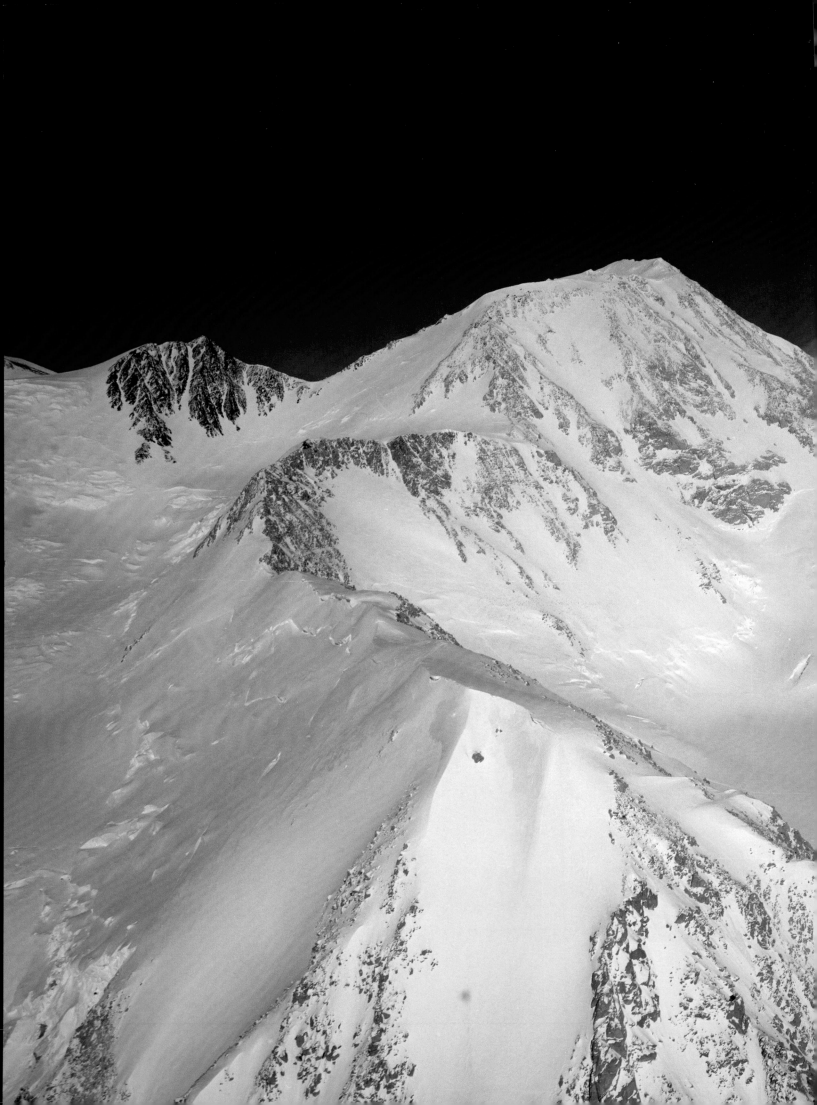

Part Three

Government and Community Organizations

Mt. McKinley.

Alaska Department of Community and Economic Development

The Trans-Alaska Pipeline.

The Alaska Department of Community and Economic Development works to promote a healthy economy and strong communities. With efficient business and occupational licensing programs, the Department strives to raise Alaska's visibility as a good place to do business. Additionally, the Department builds partnerships among Native corporations, private businesses, and communities that create new business opportunities and promote Alaska's pristine seafood products. A primary focus of the Department is Governor Frank Murkowski's strategy to link natural resource development with improved transportation infrastructure in order to provide economic opportunities in rural communities and strengthen the economy of the state. The Department will implement Governor Murkowski's strategy through an interagency rural economic development action plan called *Access to the Future*.

Alaska's modern history is based on the transportation systems built to connect the state's resource wealth with rivers and ports and to connect Alaska to the rest of the United States. In 1923, the Alaska Railroad connected the interior gold fields with coal and with the coast. Built during WWII, the Alcan Highway connected Alaska and Lower 48 states via Dawson Creek, B.C and played a key role in bolstering national security. The Alcan continues to provide a vital access and supply connection for Alaska's economy. In 1974, the Dalton Highway opened surface transportation to oil and gas development at Prudhoe Bay. The world-class lead-zinc mine at Red Dog was accessed by the DeLong Mountain Transportation System, built in 1989 to connect the mine with the port near Kivalina. Each of these improvements has produced long-term benefits, and the infrastructure has remained to build and strengthen the economy.

Today, Alaska still has vast oil, gas, mineral, and recreation resources that can be developed to provide jobs and a healthy economy. *Access to the Future* focuses on improving roads and docks to make private-sector investment in rural areas more attractive, creating more choices for well-paying, year-round jobs for rural residents. This also generates a cash economy that supplements and makes a subsistence lifestyle possible for rural residents who wish to engage in both lifestyles. Most importantly, expanded employment opportunities will give young Alaskans the option of returning home to contribute their skills and build the country for *their* children.

How will *Access to the Future* help the State of Alaska realize these goals? Governor Murkowski has taken major steps to secure legal access to resource development by streamlining the permitting process. The Governor has established "best practices" for industry to get projects moving, while not compromising "good science" environmental safeguards or sacrificing environmental quality.

In 1980, the federal government made a number of guarantees in the Alaska National Interest Lands Conservation Act to ensure that Alaska would have access to its 103 million acres of state land and be able to create an economy from them. Under the Alaska Native Claims Settlement Act, up to 44 million acres can be selected by Native regional and village corporations as well as individuals. Much of that land has had active mineral exploration. Together, the corporations and the resource companies have identified many prospects and physical access systems that become catalysts for economic activity in rural areas.

Physical access is key to economic success. Governor Murkowski and the Department of Transportation have organized the Industrial Roads Program, which is charged with the task of designing, permitting, and constructing new roads and regional ports to open Alaska's resource wealth. Connecting the Dalton Highway to Nuiqsut will open major oil and gas lease areas as well as support activity in the Northeast corner of National Petroleum Reserve-Alaska, just west of Prudhoe Bay. Some roads will connect communities and increase their economic viability. These projects will take Alaska's economy a quantum leap forward.

As part of Governor Murkowski's team, the Department of Community and Economic Development will maintain a strong focus on the basic principles of legal and physical access. Alaska has mineral, tourism, fishery, and transportation opportunities that are capable of fostering a healthy economy and strong communities.

Alaska is eager to have new people and businesses join the enjoyable effort of building in the North and creating a society that all can be proud of and an economy that will strengthen Alaska's future. Most importantly, Alaskans will welcome the next generations so they can participate and build the country for *their* children.

Edgar Blatchford, Commissioner, Alaska Department of Community and Economic Development.

The 23rd Alaska State Legislature

Outside the Alaska State Capitol building is a replica of the Liberty Bell, given to Alaska in 1950. A total of 55 full-sized replicas were cast and given as gifts to states and territories of the United States as part of a campaign for U.S. Savings Bonds.

The Alaska State Legislature has a bicameral legislature composed of two bodies, which are the House of Representatives and the Senate. The current House of Representatives consists of 27 Republicans and 13 Democrats, who were elected for a two-year term beginning January 21, 2003. Of the 40 representatives, 8 are women and 32 are men. The Senate consists of 12 Republicans and 8 Democrats of whom 4 are women and 16 are men. The 20 Senators serve four-year terms with half of the members serving staggered terms so that half are up for reelection every two years. There are two House Districts within each Senate District. The legislative branch is responsible for enacting the laws of the state and appropriating the money necessary to operate the government.

Of the 680,000 residents of Alaska, Anchorage is home to approximately half of the population of the state. The other main population centers are Fairbanks, the Matanuska-Susitna Valley, and Juneau, and each averages 35,000 residents. The rest of the state is populated with a number of towns and villages mainly accessible by water (either coastal villages or rivers which serve as ice roads in winter) or air. The logistics and costs of campaigning in House District 6 typify the challenges that state politicians face. The 10,504 registered voters live in an area which encompasses over 106,000 land miles and includes 80 communities with populations ranging from the largest, Deltana, with 1,570 residents to Chicken, Alaska with 24 residents. Double these figures for the Senator who serves twice the area, and one has an idea of the sort of campaign trail these legislators literally follow.

Vast distances between remote villages are compounded by the harsh arctic environment that covers much of the state (with winters which last seven months of the year), limited telecommunications networks, and a lack of basic services make a commitment to public service a daunting challenge. Included in this is the necessity of packing up the office and family each winter to move to Juneau, the Capitol, for the required 120 days of session, which runs from January through May.

Alaskan policymakers definitely face very unique and unusual circumstances to fulfill their obligations to their constituents.

Alaska became the 49th state in 1959. Since then, Alaska legislators have built an infrastructure of laws to secure, and transmit to succeeding generations, a heritage of political, civil and religious liberty within the Union of States. Over the years, laws were developed within a unique political environment that included complex land distribution challenges driven by the discovery of oil on Alaska's North Slope. Oil discovery accelerated the settlement of the Alaska Native Land Claims Act, unique to Alaska among all other states in America. Since statehood, Alaskans have engineered many initiatives that now serve as models for other states and nations. These include the Alaska Permanent Fund, long distance delivery of medicine, education, and arctic resource exploration and development technology, which promoted environmental technology advancements unique in the world.

State legislators have also been at the forefront of establishing regional protocols with neighboring provinces and territories in Western Canada and the Khabarovsk and Sakhalin regional legislative bodies in the Russian Far East. As North America's air crossroads to Asia and Russia, Alaskan businesses have branched out to meet global resource development challenges in these areas. This has necessitated Alaska's state policymakers to see the need to establish relations with their counterparts in nearby countries.

In 2002, the Alaska State Legislature was awarded a grant from the Library of Congress' Open World Russian Leadership Program. The grant brought 15 Russian legislators and election commission directors from five regions of the Russian Far East to Alaska to observe all aspects of the election process during the primary and general elections of 2002. Alaska's U.S. Senator Ted Stevens is honorary chairman of the Open World Russian Leadership Program. Since its creation in 1999, the program has received grants totaling $51 million from the U.S. Congress and will bring 1,600 Russian political leaders to the United States. Alaska will, again, be one of the host states.

Entrance to the Alaska State Capitol building in Juneau.

ALASKA TRAVEL INDUSTRY ASSOCIATION

The cruise-ship industry has contributed greatly to the tourism sector of Alaska's economy.

Few undertakings are as challenging and rewarding as promoting Alaska tourism. At one-fifth the size of the Lower 48 states, the "last frontier" is home to three million lakes; over 3,000 rivers; 17 of the nation's 20 highest peaks; 100,000 glaciers; and 15 national parks, preserves, and monuments.

Promoting Alaska is the responsibility of the Alaska Travel Industry Association (ATIA). ATIA is a non-profit, membership-based, trade and marketing association organized in 2001 through a merger with the Alaska Visitors Association.

The newly formed organization also assumed the marketing programs of the Alaska Tourism Marketing Counsel.

ATIA's mission is to promote travel to and within the state, develop a statewide marketing plan, increase overall awareness of the economic importance of the visitor industry, and work cooperatively with the state on tourism development and long-range planning.

ATIA's growing membership ranges from family-run bed-and-breakfasts to cruiselines. Ninety percent of ATIA's member businesses have fewer than 50 employees; 60 percent have five employees or less.

For all of Alaska's large-scale appeal, the state faces significant competition from other national destinations. In 2002, the average state tourism-marketing budget was $13.1 million; Alaska's was $7.8 million. In 2003, sixty percent of ATIA's marketing dollars came from its membership; 40 percent came from the state.

Member participation is critical to grow Alaska's national presence as a visitor destination. The combined contributions of ATIA's private members support its cooperative marketing efforts. This, in turn, allows Alaska's many smaller business operators the means to promote their businesses outside Alaska.

Though funding a national marketing program is challenging, the return on investment is considerable. Visitors spend an estimated $1,260 per person, per trip. In 2002, total direct full-year spending by visitors was estimated at $1.8 billion. The visitor industry accounts for approximately 30,700 Alaskan jobs, or one-in-eight private-sector jobs.

Because tourism remains one of the state's most important economic engines, ATIA manages a comprehensive marketing program to ensure the industry's continued growth into the new century.

Utilizing years of research, ATIA's marketing experts identify the demographics and interests of the people who visit Alaska. Guided by this research, ATIA continuously refines its marketing efforts to attract potential travelers from across the country and around the world.

Recently, ATIA adopted a new logo reflecting the opportunities and expectations this state offers to new and returning visitors: *ALASKA: Beyond your dreams. Within your reach.* SM

Public awareness is generated through advertising efforts using magazines, direct mail, television, and the Internet. ATIA continues to grow its online presence through its consumer web site.

A critical adjunct to the web site is the comprehensive Official Alaska State Vacation Planner. The Planner aids travelers in creating their ideal itinerary and identifies activities and services offered by ATIA members.

ATIA invests significant resources into public-relations efforts to generate stories that promote Alaska as a visitor destination. These efforts coincide with a vigorous government-relations program to improve conditions within the visitor industry.

The association's fulfillment department provides important information services to potential visitors and actively develops and shares potential visitor leads and planning information with its numerous statewide community partners.

Through the combined use of the Internet and its public-relations program, ATIA bolsters specific niche markets in adventure/ecotourism, sportfishing, winter, highway/Alaska Marine Highway, and cultural/historic activities and actively promotes these travel opportunities as part of its overall marketing efforts.

Creating interest in traveling to Alaska is not limited to consumers. ATIA actively participates with travel agents and tour operators to educate, identify and build alliances with groups that actively promote or provide tour opportunities within the state.

ATIA participates in high-potential international markets including Japan, German-speaking Europe, Australia, Korea, and the United Kingdom.

Tourism is the ultimate sustainable resource, and ATIA will continue to use all its resources to grow this critical industry.

A tourist favorite is a helicopter ride to a glacier. This is one example of many unique Alaskan attractions that visitors are willing to invest in for the sake of an enjoyable experience.

Alaska Industrial Development and Export Authority

AIDEA and AEA—Powering Alaska

Great growth has been achieved in Alaska's economic diversification, and the Alaska Industrial Development and Export Authority (AIDEA) is proud to have played an essential role.

Alaska has seen a more stable and efficient economy through changes and upgrades in the transportation industry and stronger communications technology. It has seen improvements in healthcare available within the state, and tourism has grown into a viable renewable resource. In line with the growth of the state, the construction and retail industries have advanced and adjusted to service the needs of Alaskans—and AIDEA has been there working hand-in-glove with the private sector every step of the way.

AIDEA is committed to providing economic growth and diversification for Alaskans and their businesses through financing assistance programs that help serve businesses statewide while providing jobs for Alaskans.

AIDEA's programs are specifically designed to help the Alaskan economy overcome two challenges: (1) assist financial institutions to finance credit-worthy business ventures in Alaska, and (2) help provide key economic infrastructure required to enable the Alaskan economy to further develop, therefore expanding the number of jobs available to Alaskans.

Functioning as a public corporation in the State of Alaska, AIDEA operates as an enterprise fund and meets all of its financial requirements, including the cost of operations and generation of additional financial assets, through its own activities. AIDEA is a profit-oriented organization, with assets totaling $1.256 billion. Because of the quality of AIDEA's assets, the organization, since 1997, has contributed dividends of over $128 million back to the State of Alaska general fund.

As Alaska's Economy Grows, So Grows the Need for Low-Cost Power

In the early 1990s, AIDEA was given oversight authority of the Alaska Energy Authority (AEA), which created a strong synergy

Ukpeagvik Inupiat Corporation, the Native village corporation in Barrow, received long-term financing for a new Alaska Commercial store in Barrow through National Bank of Alaska (now Wells Fargo Bank Alaska) and AIDEA's Loan Participation Program. AIDEA provided needed capital and helped bring new products and services to this remote community.

In 2001, AIDEA worked with local property owners to bring ownership of two of Alaska's top healthcare facilities under Alaskan ownership – North Star Hospital (shown here) and Residential Treatment Center.

that helps business ventures succeed in Alaska, while ensuring safe, reliable, and low-cost electric power is available to all Alaskans.

AEA was created by the Legislature in 1976 to provide affordable power, and therefore develop the economic welfare for all of Alaska's residents. Since its inception, AEA has made great strides toward providing safe, reliable energy to residents from Southeast to Northwest Alaska, and all points in between.

AEA places an important emphasis on lowering the costs and increasing the safety and reliability of rural power systems. Emergency responses to utility systems and fuel-storage failures are provided, as necessary, to protect the life, health, and safety of rural Alaskans.

Since AEA's inception, Alaska has experienced strong economic diversification as it moved away from heavy reliance on oil and gas development to a network of industries such as transportation, communications, healthcare and retail services.

AEA is responsible for many of the most important electric generation and distribution assets in Alaska, which includes the Bradley Lake hydroelectric facility and the Alaska Intertie that transports electricity from Willow to Healy. Alaska also has 120 independent utilities serving a total population of over 600,000, covering an immense geographic and economic landscape. It is the task of AEA to operate and maintain existing state-owned power projects to achieve the lowest reasonable consumer power costs and assist in the development of safe, reliable, and effective energy systems throughout Alaska.

AIDEA AND AEA—MEETING ALASKA'S POWER NEEDS
As an economic development engine, AIDEA helps diversify Alaska's economy and foster the growth of business. AEA helps provide safe, reliable, and efficient energy systems throughout the state that are financially viable and environmentally sound. Without access to efficient, reasonably priced power, economic development cannot proceed. Alaska's long-term economic outlook is enhanced by AIDEA and AEA combining their talents to develop and advance the general prosperity and economic welfare of the people of Alaska. This synergy is a key element in Alaska's ongoing effort to encourage new business and projects—from large to small—in rural and urban communities.

Under the Business and Export Assistance Program, AIDEA provided an 80 percent loan guarantee on a loan originated by KeyBank of Alaska to Mahay's Riverboat Service of Talkeetna. Mahay's used the loan to purchase a new, 42-foot tour boat to serve the growing tourism industry near Denali National Park.

DENALI COMMISSION

Top: Welders in Kotlik.
Above: New tank farm in Savoonga.

The Denali Commission facilitates innovative partnerships to address Alaska's infrastructure challenges, emphasizing rural communities. Since its inception through the Denali Commission Act of 1998, the Denali Commission has become a catalyst for positive change by increasing efficacy of government through collaboration. The Commission's mission is to improve the effectiveness and efficiency of government services, to develop a well-trained labor force employed in a diversified and sustainable economy, and to build and ensure the operation and maintenance of basic infrastructure.

The Denali Commission Act of 1998, sponsored by Senator Ted Stevens, designates seven appointed positions representing the major sectors of influence in Alaska. The positions include the Federal Co-Chair (federally appointed); the State Co-Chair (Governor of Alaska or appointee); University of Alaska, President; AFL-CIO, Executive President; Alaska Municipal League, President; Alaska Federation of Natives, President; and Associated General Contractors, Executive Director.

Government and Community Organizations

Old bulk fuel tanks in Savoonga.

Training

In partnership with the State of Alaska Department of Labor and Workforce Development, the Denali Commission created the Denali Training Fund to provide funding for training rural residents on careers in Construction, Operations, and Maintenance of Public Facilities. The Commission's goal of long-term sustainability is accomplished through trained residents who can operate and maintain the facilities the Commission has helped support.

Rural Energy

Most rural communities rely entirely upon diesel fuel to generate electric power. A significant number of bulk fuel storage tanks are noncompliant with state and federal safety standards, risking health problems for residents and endangering ecosystems which subsistence living depends upon. The Denali Commission provides funding to the Alaska Energy Authority, Alaska Village Electrical Cooperative, and other partners to construct safe, code-compliant, bulk-fuel-tank farms and related energy infrastructure in rural communities across Alaska.

Rural Health Care

The Denali Commission, in partnership with funding support from the U.S. Department of Health and Human Services, has designated rural health care as a top priority. Health care options are often extremely limited in rural Alaska. Medical needs or emergencies requiring hospital care frequently involve costly air transportation that can take as much time and money as a flight from New York City to Los Angeles, if weather permits. For local health care, the typical rural community clinic is aging and small. The Commission, with guidance from a Steering Committee of Rural Health Care Professionals is providing statewide support for rural health infrastructure needs.

Teller Health Clinic.

ALASKA STATE CHAMBER OF COMMERCE

The Alaska State Chamber of Commerce tells Alaska lawmakers exactly what its members require to grow their businesses and make a positive difference in the state. Membership in the State Chamber not only provides opportunities to shape rules and regulations, but also provides opportunities to build important business relationships with people representing almost every community in Alaska. In short, the Alaska State Chamber of Commerce is the voice of Alaska business.

In 1953, before Alaska even became a state, business people recognized the need to promote commerce in Alaska, so they formed the "All Alaska" Chamber of Commerce. To celebrate statehood, the name was changed to the Alaska State Chamber of Commerce, but the mission remained the same: to promote the planned, orderly growth and development of Alaska through strong private-sector business leadership that influences statewide economics and politics.

The State Chamber is the only statewide business organization with a full-time lobbyist in Juneau, and that brings a unified voice for business to the state legislature. Besides this, the State Chamber has special connections with many political leaders. The list of State Chamber members who have become involved in politics is significant, especially since these leaders bring their belief in the State Chamber's mission with them to political office. Walter Hickel, once chairman of economic development for the State Chamber, became Governor of Alaska, twice, as well as U.S. Secretary of the Interior. Former Board Chairman Bill Sheffield later became Governor of Alaska. Alaska's current Governor and long-serving U.S. Senator, Frank Murkowski, also served as chairman of the board.

Specific legislative priorities are set each year by the State Chamber's 700 business members and more than 40 local chambers of commerce around the state, the Pacific Northwest, and Canada. In the fall, the State Chamber gathers lists of Alaskan legislative priorities and positions. In November, the membership prioritizes those issues, then kicks off its promotion by meeting with every state legislator possible at its Legislative Forum in February. The State Chamber promotes its legislative priorities year-round through a strong grass-roots presence, public-awareness campaigns, and government relations.

Besides being a major conduit between business and the legislature, the State Chamber

Alaska State Chamber Members meet with the U.S. Delegation to promote building an Alaskan gas pipeline.

Alaska State Chamber Members are transported to the Green Creek Mine on Admiralty Island in Southeast Alaska by tractors.

Government and Community Organizations

provides excellent networking opportunities for its members. State Chamber members share a willingness to stand together to make a positive difference in Alaska's future. The camaraderie developed while working with the State Chamber creates a solid foundation for continued development of business relationships across the state since State Chamber members represent almost every community in Alaska.

As with any business, the State Chamber grows and changes with the times. Early on, the "All Alaska" Chamber focused on finding outside investors to help jump-start the Alaskan economy. Among its many past achievements, the State Chamber successfully lobbied for paving the Alaska Highway, launched the first Alaskan trade mission to Japan, and was instrumental in the passage of major tort-reform legislation to protect businesses from frivolous lawsuits. Currently, the State Chamber is working on changing regulations regarding the registration of lobbyists to make it easier for business to communicate concerns to the legislature.

The State Chamber's basic strategies for promoting business in the state have not changed much since its inception. The strategies adopted by the State Chamber are:

- To promote a stable, reliable economic environment in which business can prosper and to work to eliminate factors that unreasonably increase the cost and complexity of doing business in Alaska;
- To promote development and implementation of a long-term fiscal plan that addresses the diverse needs of all Alaska;

Chairman Ted Quinn dispersing Alaska: North to the Future *books to schools as part of Books to Schools, a program established in cooperation with Wyndham Publications, Inc.*

- To advocate a strong in-state education system in order to ensure a fully qualified Alaskan workforce;
- To encourage the development of Alaska's natural resources and expansion of in-state value-added processing of those resources in concert with modern principles of stewardship and sustained-yield management;
- To ensure recognition of Alaska's strategic importance to national defense and support the continued development of a strong, healthy military presence in the state;
- To support efforts to develop and promote Alaska as a foremost visitors destination.

The energy derived from all of the State Chamber members delivers the message that strong business makes a strong Alaska.

Alaska State Chamber Members inspect fish processing in Unalaska during a tour to explore industries across the state.

THE NATURE CONSERVANCY

The Nature Conservancy works in partnership with local communities, businesses, and citizens to conserve the great places of the Great Land.

The mission of The Nature Conservancy is to preserve the plants, animals, and natural communities that represent the diversity of life on Earth by protecting the lands and waters they need to survive.

The Nature Conservancy believes that taking good care of the land is a necessity. Human natural heritage is the foundation for human quality of life. That is why, for over 50 years, the Conservancy has been bringing together people of all walks of life in an effort to conserve the natural diversity of life on Earth.

People are as much a part of the landscape of Alaska as the fish and the wildlife. In order to be successful at the scale required by its mission, in order to ensure that its conservation stands the test of time, that it is durable, the Conservancy must consider human needs, both present and future, and find innovative ways to address those needs through conservation.

As the Conservancy expands its efforts to work with multiple communities, public leaders and government agencies on common threats across the large regions of Alaska, it is this commitment to people and their importance in the landscape that will provide a foundation for effective partnerships and lasting conservation action.

In order to focus its conservation resources on the most important places, the Conservancy must first determine where those places are. In other words, it needs a picture of the full range of natural diversity in the state, and it needs a big picture, one that encompasses the many ecological processes, large and small, that affect the intricate web of life in Alaska. *The Blueprint for Saving Great Places in The Great Land* publication will not only identify places necessary for long-term protection of Alaska's plants and animals, but it will also identify the threats to those species and their communities and the actions that must be taken to successfully conserve them. For the first time, a scientifically credible picture of what must be done will be available. *The Blueprint* will be available to conservationists and industry, state, and federal agencies, citizens and community leaders

Alaska's diversity of plant and animal life is a source of beauty and inspiration as well as a source of economic wealth.

Government and Community Organizations

In places like the Palmer Hay Flats, The Nature Conservancy protects both habitat and access for local hunters and other recreationists.

alike, and it will provide a common starting point for an ongoing effort to focus resources on the most important places for conservation in Alaska.

Land acquisition and conservation easements have always been important tools for conservation of habitat. In Alaska, where there are few large tracts of privately owned land, the Conservancy uses a keystone land strategy in order to ensure that its acquisitions, whatever their size may be, have a landscape-scale impact. Keystone lands are those places on which the ecological health of very large landscapes depend. These lands do not have to be large themselves—they can be 1,000 acres or 100 acres or only 5 acres—but their impact is immense. It may be that the land is a critical feeding ground where brown bear regularly appear to gorge themselves on sockeye salmon. It could be that it is a migratory stop where thousands of birds rest on their long journey to their feeding grounds. Or perhaps it is the only piece of land for sale among tens of thousands of acres of protected, critical habitat. Maybe it is the gateway to a pristine wilderness area. The "keystone impact" of protecting such lands can affect tens, even hundreds of thousands of acres and abate one of the most significant threats to Alaska's parks and refuges: incompatible development of private inholdings.

The Conservancy and its members have been responsible for the protection of more than 12 million acres in 50 states and Canada. It has helped like-minded partner organizations to preserve millions of acres in Latin America, the Caribbean, the Pacific, and Asia. Although some Conservancy-acquired areas are sold for management to other conservation groups, both public and private, the Conservancy owns more than 1,400 preserves—the largest private system of nature sanctuaries in the world.

The Nature Conservancy opened its field office in Alaska in 1988. To date, The Nature Conservancy has protected tens of thousands of acres of critical habitat in the state by working with partners, both public and private, to ensure the balance between economic needs and ecological values.

By conserving habitat now, the Conservancy protects plant and animal life for future generations.

Committee on Economic Development, International Trade and Tourism

Economic development, trade, and tourism are areas of vital importance to the State of Alaska. Alaska is a land of vast natural resources situated uniquely at the top of the world with ready access by air and sea to North America, Europe, Asia, and Russia. Trade between what is now Alaska and other parts of the world has been conducted for centuries. Early trade of timber, fish, and fur was followed by discoveries of gold and other precious metals. Discoveries of oil in the early part of this century have vastly increased the growth and development of Alaska. More recently, tourism has developed into one of the primary industries within Alaska.

Economic development, international trade, and tourism are of such immense importance to the State of Alaska that a special committee within the legislature has been developed to promote and assist in these issues. The Committee on Economic Development, International Trade and Tourism is tasked with developing and maintaining relationships with worldwide communities. Long-time trade partners, Asia, Canada, and the Russian Far East, need to be maintained while new partnerships are developed. Promotion of the tourist industry is a constant theme. The legislature can and should assist local business in the promotion of these areas of vital interest to the state.

Representative Cheryll Heinze, a long-time Alaskan, has been chosen to chair the Special Committee on Economic Development, International Trade and Tourism. Representative Heinze is uniquely suited to her role. She has an extensive history of world travel, having lived abroad for many years. She served as special projects coordinator in the Division of Tourism and later as a fishing-lodge owner and manager; experiences which have allowed her first-hand exposure to many areas of interest to the state as a whole. As committee Chair, she is committed to maintaining current relationships while aggressively seeking new opportunities.

Speaker of the House Pete Kott is another key member of the committee. Other members include Representative Lesil McGuire who serves as vice-chair. Representative McGuire chaired the committee last session. Representative Vic Kohring, Representative Nancy Dahlstrom, Representative Sharon Cissna, and Representative Harry Crawford round out the remainder of the committee positions.

It may seem a bit crowded, but these anglers are all actually catching fish! The salmon are running and everyone is getting a bite. Anglers travelling to Alaska can fish with plenty of company, or if they prefer, they can take advantage of the more remote destinations. Charter flights are available to some of the most beautiful and pristine fishing locales on the planet.

The crab pots are stacked and ready for the next trip. The seafood industry has long been a staple of Alaska's economy and many of the coastal communities thrive because of the state's care in fisheries management.

Resource Development Council

Resource development is Alaska's economic foundation and a critical component of the nation's economy. While estimates vary according to the source, Alaska may hold 30 percent of the country's oil reserves and nearly 20 percent of its natural gas. Its coal resources could power America for centuries, and other mineral deposits such as zinc and gold are world class. Meanwhile, Alaska's fishery resources are second to none, and the state's natural beauty and frontier character has made it a major international visitor destination.

With such a rich endowment of natural resources comes great potential and opportunity for Alaska and its economy. The Resource Development Council for Alaska, Inc. (RDC) is where Alaskans from all walks of life come together to work for the sensible and progressive development of Alaska's abundant natural resources.

RDC is a statewide, non-profit, membership-funded organization made up of individuals, local communities, organized labor, Native corporations, and businesses from all resource sectors. Through RDC these interests work together to promote and support responsible development of Alaska's resources.

RDC was formed in 1975, originally as the Organization for Management of Alaska's Resources (OMAR). After three years working to obtain a Trans-Alaska gas pipeline, RDC changed its name to reflect its broader agenda of education and advocacy on all resource issues in Alaska.

RDC advocates for a reliable and economical long-term federal and state timber harvest.

RDC's Mission Statement
- Growing Alaska Through Responsible Resource Development.

RDC's Goals
- Promote sound resource development in Alaska.
- Link diverse interests on resource issues.
- Sustain and expand a diverse membership.
- Educate the public, policymakers, and students on resource issues.

RDC works for all resource sectors, including mining, oil and gas, fisheries, timber, and tourism. RDC provides forums for policy debate and analysis to help guide Alaska in these areas, as well as in land use, transportation, power development, international trade, and economic development.

RDC supports initiatives to encourage new exploration and development of Alaska's oil, gas, and mineral deposits, as well as increased production from existing deposits. Pictured is a miner pouring a gold bar at the Ft. Knox Mine.

Government and Community Organizations

The Alaska Nurses Association

A moment of rest for a busy itinerant nurse.

The Alaska Nurses Association (AaNA) is the professional association for Registered Nurses (RNs) in the State of Alaska and is a constituent member of the American Nurses Association. The Association's mission is to work for the improvement of the health standards and availability of health services for all people, foster high standards for nursing, stimulate and promote the professional development of nurses, and advance their economic and general welfare.

AaNA was officially formed in May 1952 and incorporated as a nonprofit in 1953, although groups such as the Ketchikan Graduate Nurses Club and the Fairbanks Nurses Association had been active politically in the Territory since 1937. It was the Fairbanks nurses group that first affiliated with the American Nurses Association in 1940.

AaNA pursues its mission in Alaska by facilitating work in several areas. The Association lobbies on legislation that is important to RNs and the patients that they serve. AaNA also works to maintain a highly qualified nursing workforce by certifying continuing education offerings, promoting educational offerings, and supporting coalitions, such as the Alaska Colleagues in Caring, to advance nursing. AaNA monitors trends in the health care system, activities of the Board of Nursing and the Board of Medicine, and reports that affect RNs' practice. The Association also advocates for changes to advance their practice.

AaNA also advocated for improvements and high quality in the workplace for Registered Nurses. Since 1994, the AaNA has grown to represent RNs in three bargaining units: Providence Registered Nurses Bargaining Unit, Central Peninsula General Hospital RNs United Bargaining Unit, and Ketchikan General Hospital RNs United Bargaining Unit. Serving as the union for these RNs, AaNA is able to negotiate contracts that advance competitive wages and working conditions that attract and retain nurses, and promote the best in nursing practice. In addition, the Association's bargaining units have initiated work on issues, such as health and safety, which AaNA has then been able to share with RNs throughout the Alaska.

AaNA—To Protect and Advance the Practice of Nursing in Alaska.

Associated General Contractors of Alaska

Bartlett Memorial project.

Established in 1948, the Associated General Contractors of Alaska (AGC) is the undisputed champion of the state's construction industry. The organization is one of 100 chapters of the widely acclaimed Associated General Contractors of America, the oldest construction association in the nation.

AGC members come from all realms of Alaska's construction industry, and membership benefits apply to every employee, not just administrative echelons. Members can tap into a vast spectrum of services including workforce education and training; safety and environmental expertise; and assistance in labor, legal, legislative, and information arenas.

Following a tradition spanning more than five decades, AGC is a voice for the current construction establishment.

But AGC has an eye on the future too. Workers who helped build the Trans-Alaskan oil pipeline in the 1970s during the state's last building boom are now at least 45 years old. The trade is facing a critical shortage of workers at the same time demand for services is growing. To meet the challenge, AGC is promoting the notion that not everyone needs to go to college to excel as well-paid professionals.

For many in Alaska's Interior, construction is truly the OTHER four-year degree. The drawing factor? Straight out of high school, apprentices can earn as they learn, eventually collecting paychecks that can be double a college graduate's income—minus the burden of paying off hefty student loans.

Given Alaska's infinite chance for growth, AGC is going for an even younger audience through *Construction Futures*. The multimedia program is developed by the Associated General Contractors of America and Scholastic Inc. to promote the construction industry to secondary students. *Build Up!* is aimed at the elementary-school level and features hands-on projects. *On Site!* is designed for grades six through nine and is geared for use in math and social-studies subjects, with applications in language arts and sciences.

A third program, *Core Curriculum*, introduces participating high school students to construction basics including safety, math, and basic equipment applications.

What the future boils down to, AGC believes, is sharing the rewarding options available in building the Next Frontier.

AGC members come from all areas of Alaska's construction industry.

Anchorage Chamber of Commerce

Lieutenant General Norton Schwartz, Commander of the Alaskan Command, and Anchorage Chamber of Commerce Chair Eric Britten pose for photos during the 2002 Military Appreciation Luncheon honoring the nation's service men and women.

Like many things Alaskan, the city of Anchorage is as big and as diverse as the state itself. The municipal boundaries stretch more than 50 miles, from Portage Glacier to the head of Cook Inlet, encompassing 1,955 square miles—about the same size as the state of Delaware. The city's diversity is exemplified by its demographic make-up: 30 percent of the population consists of different minorities and up to 85 languages are spoken in the school district. American Indians and Alaska Natives make up 7.9 percent of the city's population.

Anchorage is also a city of extremes. Long summer days encourage brilliant flower gardens, while dark winter nights are made magical with the reflection of the northern lights. Moose and bear are common wanderers through neighborhoods, and phrases like "termination dust" and "breakup" take on whole new meanings. Those with an adventurous spirit call Anchorage home.

The Anchorage Chamber of Commerce is reflective of the state's uniqueness. With more than 1,200 members representing 55,000 employees, it is an independent non-profit membership organization. Established in 1915, when Anchorage was still a "tent city," the Chamber has a long history of business advocacy and community service. Advocacy continues to remain a focus as well as economic and business development, and member services.

The Anchorage Chamber has been instrumental in creating a business-led coalition—Vision Anchorage—to address the need for diversifying and growing Anchorage's economic base. While the coalition faces the difficult tasks of implementation of strategies and achievement of goals, the Anchorage Chamber is committed to putting all of its energy, enthusiasm, and resources towards ensuring the success of such a momentous undertaking.

The people of the community have always looked to the Anchorage Chamber when things need to be done. For example, City-Wide Clean Up occurs every spring. Thousands of volunteers collect and dispose of a million pounds of trash in order to spruce up the city for the summer. During Military Appreciation Week, the Military Committee hosts thousands of military personnel at picnics held simultaneously at Elmendorf Air Force Base and Fort Richardson. City of Lights is a program that encourages businesses and residents to hang white miniature lights to enhance Anchorage's beauty from October through March.

The Anchorage Chamber of Commerce has played a vital role in the growth and development of Anchorage and will continue to be a leader in the years ahead.

The kick-off to the 34-year tradition of the Anchorage Chamber's City Wide Clean-Up is the Blue Jeans Luncheon featuring a "Trashy Fashion Contest."

Participants from the business community work together to create the "trashiest" song and dance routine.

EXPORT COUNCIL OF ALASKA

Members of the Export Council of Alaska, left to right: Joe Henri, Treasurer; Ron Sheardown; Steve Smirnoff; Lee Wareham; Lloyd Morris; John Norman; Chuck Webber, Chairman; Bill Noll; Elary Gromhoff; Roger Chan; Steve Grabacki; Bob Stiles; Tony Follett, Vice Chairman; Tina Fleckenstein, Staff; Chuck Becker (seated), Secretary.

Free and fair trade is America's foil in the international arena. And, free and fair trade is essential to Alaska's economic growth and prosperity. During the decade of the nineties, the value of merchandise exports alone ran right at ten percent of gross state product at $2.5 billion. Now, the export of services by Alaskans has begun to grow rapidly with major penetrations into the markets of Canada and the Russian Far East.

Thousands of Alaskans work at jobs that depend on exports, including jobs in the seafood industry, mining, oilfield service, and other sectors. These export-dependent jobs pay wages an estimated 13 to 18 percent higher than average. In 2001, Alaska's worldwide commodity exports again hit $2.5 billion. Exports bring new money into the economy of the state and its towns and villages.

The global marketplace has undergone dramatic and fundamental change since the end of the Cold War. With American leadership, first in the General Agreement on Tariffs and Trade (GATT) and the successor, the World Trade Organization, governments have significantly reduced barriers to trade and investment. The flow of goods, services, and investment capital around the world has exploded over the past decade. And Alaska is a direct beneficiary.

In 1993, Alaskan exports to Canada were valued at $74 million. At the end of 1994, the value climbed to $106 million before rocketing to $210 million in 1995. What made the difference? One key development occurred on January 1, 1994, when the provisions negotiated under the North American Free Trade Agreement became effective.

The Export Council of Alaska is a 25-member body, all of whom are appointed by the Secretary of the U.S. Department of Commerce to serve as his advisors on trade issues; to act as advisors to the Alaska Export Assistance Center, the field unit of the U.S. Commercial Service; and to highlight the benefits of exporting for the Alaska business community. Members of the Council are dedicated to ensuring that Alaska's exports continue to grow.

Policy Committee Members meeting with Governor Frank Murkowski (center), left to right: John Norman, Joe Henri, Lee Wareham, Steve Smirnoff, Lloyd Morris, Ron Sheardown, Chuck Webber, Governor Frank Murkowski, Tina Fleckenstein, Bill Noll, Chuck Becket, Tony Follett, Roger Chan.

Alaska Department of Military & Veterans Affairs

Proudly welcoming loved ones home.

The mission of the Alaska Department of Military & Veterans Affairs is to provide ready, relevant, reliable military forces capable of rapid deployment in response to emergencies and disasters.

With more than 5,000 employees on a full- and part-time basis, it is an easy blend of civilians and military members. The Department operates armories and facilities at 76 locations, including Kulis ANG Base, Elmendorf AFB, and Eielson AFB. They fly 62 aircraft.

In the aftermath of September 11, the Alaska National Guard deployed nearly 300 soldiers and airmen to 19 airports across Alaska to increase security as part of Operation Noble Eagle. Along with several dozen members of the Alaska State Defense Force, they also sent security teams to several key checkpoints along the 800-mile-long Trans-Alaska oil pipeline.

More than 700 airmen and soldiers were deployed to overseas locations in direct support of Operation Enduring Freedom. Alaska Air Guard tankers refueled forces transiting the Pacific.

Alaska's Office of Homeland Security coordinates with local, state, and federal agencies to assure the safety and security of citizens.

The Division of Emergency Services' teams have substantial daily impact in the lives of Alaskans. Along with a broad range of portfolios that include mitigation issues associated with earthquakes, tsunamis, floods, fires, avalanches and other disasters, they also engage in the detailed work of disaster recovery. The State Emergency Coordination Center provides 24/7 continuity.

The Guard's Rescue Coordination Center directs more than 300 missions annually resulting in Guard rescue teams saving or assisting about 150 people.

Alaska's Military Youth Academy graduates about 100 cadets, previously tagged as youth-at-risk, from each of two 22-week classes annually. About 85 percent of the youngsters go on to college or vocational school afterward.

The Guard's Counter-Drug Program, responsible for assisting law enforcement agencies in removing illegal drugs, recently helped capture nearly $28 million in marijuana, cocaine, and other illicit drugs from the streets. They assisted officers in seizing $881,000 in cash and 122 weapons.

The Army Guard continues preparations to operate America's Missile Defense program. New facilities, new specialties, and new opportunities are fueling the transformation.

In direct support of Operation Enduring Freedom, the Alaska Air Guard refueled forces transiting the Pacific.

The U.S. Fish and Wildlife Service

Camping on Kigul Island.

More than a thousand decades ago, when the last great ice age flooded the earth in a frozen ocean, the world's wildlife discovered its ark in Alaska, a relatively ice-free oasis. Wooly mammoths, saber-toothed tigers, wild horses, caribou, and a host of other creatures of the land, water, and air—the denizens of today's Great Land—migrated across the Bering Land Bridge to a new world. Following this rich, self-renewing, living resource came the first humans to inhabit the New World, the nomadic hunters who became the ancestors of today's Alaska Natives.

The wildlife they sought gave them not only food, shelter, and clothing, but also a spiritual link between themselves and the complex and mysterious, interdependent elements of nature.

Today, Alaska continues to embrace the world's finest, richest, and most vibrant wildlife community. Alaska's new-age nomads—tourists—travel from every corner of the globe to enjoy the state's wildlife treasures. Alaskans continue to depend on fish and wildlife for subsistence, commerce, quality of life, and as vital parts of the ecosystems upon which everyone depends. But Alaska's rapidly growing population and economy place increasing demands on its fragile, boreal environment.

The U.S. Fish and Wildlife Service manages the 95-million-acre National Wildlife Refuge System, which encompasses 542 national wildlife refuges, thousands of small wetlands, and other special-management areas across the nation. More than 82 percent of this acreage is in Alaska. The Service also enforces federal wildlife laws, administers the Endangered Species Act, manages migratory bird populations, restores nationally significant fisheries, conserves and restores wildlife habitat such as wetlands, and helps foreign governments with their conservation efforts. Finally, the Service oversees the Federal Aid program that distributes hundreds of millions of dollars in excise taxes on fishing, hunting, and boating equipment to state fish and wildlife agencies. In addition to being a predominately biological Federal agency, the Service has also become an identifier of obstacles and opportunities, a facilitator of public forums, and a seeker of collaborative solutions. The U.S. Fish and Wildlife Service mission is: "…to conserve, protect, and enhance fish, wildlife, plants, and their habitats for the continuing benefit of the American people."

Caribou.

Team Alaska

AeroMap U.S.
2014 Merrill Field Drive
Anchorage, Alaska 99501
Telephone: 907.272.4495
Facsimile: 907.274.3265
Web Site: www.aeromap.com
Page 204

Alaska Aerospace Development Corporation
4300 B Street, Suite 101
Anchorage, Alaska 99503
Telephone: 907.561.3338
Facsimile: 907.561.3339
Web Site: www.akaerospace.com
Page 174

Alaska Center for the Performing Arts, Inc.
621 West 6th Avenue
Anchorage, Alaska 99501
Telephone: 907.263.2900
Facsimile: 907.263.2927
Web Site: www.alaskapac.org
Page 249

Alaska Department of Community and Economic Development
P. O. Box 110800
Juneau, Alaska 99811
Telephone: 907.465.2500
Facsimile: 907.465.5442
Web Site: www.dced.state.ak.us
Page 254

Alaska Department of Military & Veterans Affairs
P.O. Box 5800, Camp Denali
Fort Richardson, Alaska 99505-5800
Telephone: 907.428.6031
Facsimile: 907.428.6035
Web Site: www.ngchak.org/dmva
Page 274

Alaska Health Resources
3001 Madison Way
Anchorage, Alaska 99508
Telephone: 907.272.2323
Facsimile: 907.279.1507
Web Site: www.akres.com
Page 239

The Alaska Industrial Development & Export Authority
813 West Northern Lights Boulevard
Anchorage, Alaska 99503
Telephone: 907.269.3000
Facsimile: 907.269.3044
Web Site: www.aidea.org
Page 260

Alaska Interstate Construction, LLC
P. O. Box 233769
Anchorage, Alaska 99523
Telephone: 907.562.2792
Facsimile: 907.562.4179
Web Site: www.aicllc.com
Page 218

Alaska Native Tribal Health Consortium
4141 Ambassador Drive
Anchorage, Alaska 99508
Telephone: 907.729.1900
Facsimile: 907.729.1901
Web Site: www.anthc.org
Page 228

Alaska Nurses Association
2207 East Tudor, Suite 34
Anchorage, Alaska 99507-1069
Telephone: 907.274.0827
Facsimile: 907.272.0292
Web Site: www.aknurse.org
Page 270

Alaska Option Services Corporation
P. O. Box 196233
Anchorage, Alaska 99519-6233
Telephone: 907.563.0078
Web Site: www.alaskaoption.com
Page 197

Alaska Regional Hospital
2801 DeBarr Road
Anchorage, Alaska 99508
Telephone: 907.276.1131
Web Site: www.alaskaregional.com
Page 238

Alaska's Marine Highway
3132 Channel Drive
Juneau, Alaska 99801-7898
Telephone: 907.465.3941
Facsimile: 907.277.4829
Web Site: www.alaska.gov/ferry
Page 157

Alaska's Regional Port, the Port of Anchorage
2000 Anchorage Port Road
Anchorage, Alaska 99501
Telephone: 907.343.6200
Facsimile: 907.277.5636
Web Site: www.ci.anchorage.ak.us
Page 168

Alaska State Chamber of Commerce
217 2nd Street, Suite 201
Juneau, Alaska 99801
Telephone: 907.586.2323
Facsimile: 907.463.5515
Web Site: www.alaskachamber.com
Page 264

Alaska State Legislature State Capitol
Juneau, Alaska 99801
Web Site: www.legis.state.ak.us
Page 256

Alaska Travel Industry Association
2600 Cordova Street, Suite 201
Anchorage, Alaska 99503
Telephone: 907. 929.2842
Facsimile: 907. 561.5727
Web Site: www.alaskatia.com
Page 258

ALASKA USA FEDERAL
CREDIT UNION
P. O. Box 196613
Anchorage, Alaska 99516-6613
Telephone: 907.563.4567
Web Site: www.alaskausa.org
Page 194

AMERICAN SEAFOODS GROUP
Market Place Tower
2025 First Avenue, Suite 1200
Seattle, Washington 98121
Telephone: 206.374.1515
Facsimile: 206.374.1516
Web Site:
www.americanseafoods.com
Page 178

ANCHORAGE CHAMBER OF
COMMERCE
441 West Fifth Avenue,
Suite 300
Anchorage, Alaska 99501
Telephone: 907.677.7105
Facsimile: 907.272.4117
Web Site:
www.anchoragechamber.org
Page 272

ANCHORAGE CHRYSLER
DODGE CENTER
2501 East 5th Avenue
Anchorage, Alaska 99501
Telephone: 907.276.1331
Facsimile: 907.276.4191
Web Site:
www.Alaskacarnet.com
Page 242

ANCHORAGE FUR RENDEZVOUS
400 D Street, #200
Anchorage, Alaska 99501
Telephone: 907.274.1177
Facsimile: 907.274.2199
Web Site: www.furrondy.net
Page 251

ANCHORAGE SCHOOL DISTRICT
P. O. Box 196614
Anchorage, Alaska 99519-6614
Telephone: 907.742.4000
Page 237

ARCTIC SLOPE REGIONAL
CORPORATION
301 Arctic Slope Avenue,
Suite 300
Anchorage, Alaska 99518
Telephone: 907. 339.6000
Facsimile: 907. 349.5476
Web Site: www.asrc.com
Page 208

ASPEN HOTELS OF ALASKA
P. O. Box 100860
Anchorage, Alaska 99510
Telephone: 907.274.2151
Facsimile: 907.274.2152
Web Site:
www.aspenhotelsak.com
Page 250

ASSOCIATED GENERAL
CONTRACTORS OF ALASKA
8005 Schoon Street
Anchorage, Alaska 99518
Telephone: 907.561.5354
Facsimile: 907.562.6118
Page 271

AVIS RENT A CAR
P. O. Box 190028
Anchorage, Alaska 99519-0028
Telephone: 907.243.4300
Facsimile: 907.249.8247
Web Site: www.avisalaska.com
Page 170

CARLILE TRANSPORTATION
SYSTEMS
35225 Enchanted
Parkway South
Federal Way, Washington 98003
Telephone: 253.874.2633
Facsimile: 253.874.8615
Web Site: www.carlilekw.com
Page 172

CHUGACH ELECTRIC
ASSOCIATION
P. O. Box 196300
Anchorage, Alaska 99519-6300
Telephone: 907.563.7494
Facsimile: 907.562.0027
Web Site:
www.chugachelectric.com
Page 162

CITY AND BOROUGH OF JUNEAU
155 South Seward Street
Juneau, Alaska 99801
Telephone: 907.586.5240
Facsimile: 907.586.5385
Web Site: www.juneau.org
Page 136

CITY OF SEWARD
P. O. Box 167
Seward, Alaska 99664
Telephone: 907.224.4047
Facsimile: 907.224.4038
Web Site: www.cityofseward.net
Page 146

CITY AND BOROUGH OF SITKA
100 Lincoln Street
Sitka, Alaska 99835
Telephone: 907.747.3294
Facsimile: 907.747.7403
Web Site: www.cityofsitka.com
Page 130

CITY OF WRANGELL
P. O. Box 531
Wrangell, Alaska 99929
Telephone: 907.874.2381
Facsimile: 907.874.3952
Web Site: www.wrangell.com
Page 144

COMMITTEE ON ECONOMIC
DEVELOPMENT, INTERNATIONAL
TRADE AND TOURISM
State Capitol, Room 416
Juneau, Alaska 99801
Telephone: 907.465.4930
Facsimile: 907.465.3834
Page 268

CROWLEY ALASKA
2525 C Street
Anchorage, Alaska 99503
Telephone: 907.278.4978
Page 154

DELANEY, WILES, HAYES, GERETY, ELLIS & YOUNG, INC.
1007 West 3rd Avenue, Suite 400
Anchorage, Alaska 99501
Telephone: 907.279.3581
Facsimile: 907.277.1331
Web Site: www.delaneywiles.com
Page 190

DENALI COMMISSION
510 L Street, Suite 410
Anchorage, Alaska 99501
Telephone: 907.271.1414
Facsimile: 907.271.1415
Web Site: www.denali.gov
Page 262

DOWLAND BACH CORP.
P. O. Box 230126
Anchorage, Alaska 99523-0126
Telephone: 907.562.5818
Facsimile: 907.562.5816
Web Site: www.dowlandbach.com
Page 222

EXPORT COUNCIL OF ALASKA
550 West Seventh Avenue, Suite 1770
Anchorage, Alaska 99501
Telephone: 907.271.6237
Facsimile: 907.271.6242
Web Site: www.alaska.net\~export
Page 273

FAIRBANKS INTERNATIONAL AIRPORT
6450 Airport Way
Fairbanks, Alaska 99709
Telephone: 907.474.2500
Facsimile: 907.474.2513
Web Site: www.dot.state.ak.us/faiiap/
Page 164

FEDEX EXPRESS
6050 Rockwell Avenue
Anchorage, Alaska 99502
Telephone: 907.249.3181
Facsimile: 907.249.3178
Web Site: www.fedex.com
Page 166

GCI
2550 Denali, Suite 1000
Anchorage, Alaska 99503
Telephone: 907.265.5600
Facsimile: 907.265.5676
Web Site: www.gci.com
Page 158

GOLDEN VALLEY ELECTRIC ASSOCIATION, INC.
P. O. Box 71249
Fairbanks, Alaska 99707
Telephone: 907.452.1151
Facsimile: 907.451.5633
Web Site: www.gvea.com
Page 150

H. C. PRICE CO.
301 West Northern Lights Boulevard, Suite 300
Anchorage, Alaska 99503
Telephone: 907.278.4400
Facsimile: 907.278.3255
Page 216

HISTORIC ANCHORAGE HOTEL
330 E Street
Anchorage, Alaska 99501
Telephone: 907.272.4553
Facsimile: 907.277.4483
Web Site: www.historicanchoragehotel.com
Page 248

HORIZON LINES OF ALASKA, LLC
1717 Tidewater Road
Anchorage, Alaska 99501
Telephone: 907.263.5611
Facsimile: 907.263.5620
Web Site: www.horizon-lines.com
Page 175

HUNA TOTEM CORPORATION
9301 Glacier Highway
Juneau, Alaska 99801-9306
Telephone: 907.789.1773
Facsimile: 907.789.1896
Web Site: www.hunatotem.com
Page 173

METLAKATLA INDIAN COMMUNITY
Box 8
Metlakatla, Alaska 99926
Telephone: 907.886.4441
Facsimile: 907.886.4470
Page 140

THE NATURE CONSERVANCY IN ALASKA
412 West First Avenue, Suite 200
Anchorage, Alaska 99501
Telephone: 907.276.3133
Facsimile: 907.276.2584
Web Site: nature.org/alaska
Page 266

N. C. MACHINERY CO.
P. O. Box 88486
Seattle, Washington 98138
Telephone: 425.251.5800
Page 220

PEAK OILFIELD SERVICE CO.
2525 C Street, Suite 201
Anchorage, Alaska 99503
Telephone: 907.263.7000
Facsimile: 907.263.7070
Web Site: www.peakalaska.com
Page 186

PGS ONSHORE, INC.
341 West Tudor Road, Suite 206
Anchorage, Alaska 99503
Telephone: 907. 569.4049
Facsimile: 907. 569.4047
Web Site: www.pgs.com
Page 184

PORT OF ANCHORAGE
2000 Anchorage Port Road
Anchorage, Alaska 99501
Telephone: 907.343.6200
Facsimile: 907.277.5636
Web Site: www.ci.anchorage.ak.us
Page 168

PREMERA BLUE CROSS BLUE
SHIELD OF ALASKA
2550 Denali Street, Suite 1404
Anchorage, Alaska 99503
Telephone: 907.258.5065
Facsimile: 907.258.1619
Web Site: www.premera.com
Page 198

PROVIDENCE HEALTH SYSTEM
IN ALASKA
P. O. Box 196609
Anchorage, Alaska 99519
Telephone: 907.261.5692
Facsimile: 907.261.2038
Web Site:
www.providence.org/alaska
Page 234

RESOURCE DEVELOPMENT
COUNCIL FOR ALASKA, INC.
121 West Fireweed Lane,
Suite 250
Anchorage, Alaska 99503
Telephone: 907.276.0700
Facsimile: 907.276.3887
Web Site: www.akrdc.org
Page 269

SEATTLE MORTGAGE (ALASKA)
4300 B Street, Suite 206
Anchorage, Alaska 99503
Telephone:
907.562.LOAN (5626)
Facsimile: 907.562.7798
Page 201

SUPERIOR PLBG. & HTG. INC.
8861 Elim Street
Anchorage, Alaska 99507
Telephone: 907.349.6572
Facsimile: 907.349.4480
Web Site:
www.superiorpandh.com
Page 224

TECK COMINCO ALASKA
INCORPORATED
3105 Lakeshore Drive,
Building A, Suite 101
Anchorage, Alaska 99517
Telephone: 907.266.4567
Facsimile: 907.266.4568
Web Site:
www.teck.com/operations/
reddog/index.html
Page 182

TED STEVENS ANCHORAGE
INTERNATIONAL AIRPORT
P. O. Box 196960
Anchorage, Alaska 99519-6960
Telephone: 907.266.2525
Facsimile: 907.266.2458
Web Site:
www.anchorageairport.com
Page 160

U.S. FISH AND
WILDLIFE SERVICE
1011 East Tudor Road
Anchorage, Alaska 99503
Telephone: 907.786.3309
Facsimile: 907.786.3495
Web Site: http://alaska.fws.gov
Page 275

VECO CORPORATION
3601 C Street
Anchorage, Alaska 99503
Telephone: 907.264.8100
Facsimile: 907.264.8130
Web Site: www.veco.com
Page 214

WELLS FARGO BANK ALASKA
301 West Northern
Lights Boulevard
Anchorage, Alaska 99503
Telephone: 1.800.TO.WELLS
Facsimile: 907.265.2043
Web Site: www.wellsfargo.com
Page 202

ADDITIONAL PROJECT
SUPPORT BY:
Alaska Executive Search, Inc.
Anchorage, Alaska

BIBLIOGRAPHY

Alaska Bureau of Vital Statistics

Alaska Department of Commerce & Economic Development

Alaska Department of Labor, Research and Analysis

Alaska Department of Revenue

Alaska Place Names, Fourth Edition, Alan Edward Schorr

Alaska Travel Smart, John Muir Publications

http://216.239.53.100/search?q=cache:4wombDdLutsC:crm.cr.nps.gov/archive/22-10/22-10-14.pdf+alaska+tourists+1900s&hl=en&ie=UTF-8

http://almis.labor.state.ak.us

http://arcticcircle.uconn.edu/SEEJ/Landclaims/ancsa1.html

http://encarta.msn.com/encncnet/refpages/refarticle.aspx?refid=761568687

http://excursia.bestreadguide.com/juneau/stories/19990616/fea_nativepeople.shtml

http://litsite.alaska.edu/uaa/akmaps/farnorth.html

http://litsite.alaska.edu/uaa/aktraditions/ancsa.html

http://memory.loc.gov/intldl/mtfhtml/mfak/mfaknative.html

http://sled.alaska.edu/akfaq/akchron.html

http://sled.alaska.edu/akfaq/aksymb.html

http://volcano.und.nodak.edu/vwdocs/current_volcs/alaska/akutan.html

http://volcano.und.nodak.edu/vwdocs/current_volcs/shishaldin/shishaldin.html

http://volcano.und.nodak.edu/vwdocs/volc_images/north_america/alaska/isanotski.html

http://www.aerostates.org/html/activity/AK_space.html

http://www.akaerospace.com/frames1.html

http://www.alaskaalliance.org/villages/akutan.htm

http://www.alaska.com/akcom/southcentral/cities/v-akcom/story

http://www.alaskan.com

http://www.alaskaone.com

http://www.alaska.or.kr/eng/2-trade-fisheries-e.htm

http://www.alaskarailroad.com/corporate/FactSheet.html

http://www.alaskasbest.com/facts.htm

http://www.alaskatobycharter.com/TourI-Palmer.htm

http://www.alaskaunited.com/descript.htm

http://www.american.edu/TED/Eskimo.htm

http://www.angelfire.com/ak/alaskainfo/akhist.html

http://www.avo.alaska.edu/avo4/atlas/Okmok.htm

http://www.britannica.com/eb/article?eu=1788

http://www.cabelasiditarod.com/2001/prerace.html

http://www.commonwealthnorth.org/studygroup/timeline.html

http://www.commonwealthnorth.org/transcripts/burden2000/

http://www.dced.state.ak.us/tourism.student.htm

http://www.dnr.state.ak.us/

http://www.dot.state.ak.us/

http://www.explorenorth.com/library/communities/alaska/bl-Kipnuk.htm

http://www.explorenorth.com/native-ak.html

http://www.fishingforthefuture.org

http://www.gci.com/about/press/au_laid.htm

http://www.geocities.com/Athens/Acropolis/4870/AlaskaHistory.html

http://www.geocities.com/Heartland/Bluffs/8336/yukon_kate.html

http://www.geocities.com/TheTropics/4363/alaska/ak_geo.html

http://www.gi.alaska.edu/InfoOffice/gorge.html

http://www.gi.alaska.edu/ScienceForum/ASFO/091.html

http://www.glaciers/quickfacts.html

http://www.iditarod.com/sled_dog_race.html

http://www.inalaska.com/d/seward/economy.html

http://www.kokogiak.com/klon/soapy.html

http://www.library.state.ak.us/goldrush/LEGACY/pop.htm

http://www.margaretdeefholts.com/soapysmith.html

http://www.mysteriesofcanada.com/Yukon/klondike_kate.htm

http://www.nativefederation.org/frames/calendar.html

http://www.nomealaska.org/vc/iditarod.htm

http://www.nps.gov/dena/

http://www.nps.gov/yuch/Expanded/site_bulletins/yukon_river_float/yukon_river.htm

http://www.oceania.org.au/soundnet/may02/bow.html

http://www.peakbagger.com/peak/mckinle.htm

http://www.qiviut.com

http://www.sled.alaska.edu/akfaq/akchron.html

http://www.slowfood.com/img_sito/riviste/new_slow/EN/34/sourdoughs.html

http://www.southwestalaska.com/

http://www.state.ak.us/

http://www.state.ak.us/adfg/subsist

http://www.uaf.edu/asgp/affmain.htm

http://www.uaf.edu/seagrant/NewsMedia/98ASJ/06.04.98_MuskOs.html

http://www.valdezcampgrounds.com/akinfo.html

http://www.welcometoalaska.com/facts.htm

http://www.wff.nasa.gov

http://www.whaleroute.com/migrate

Scott Goldsmith, ISER Data Base and Economic Projections

The Alaska Almanac, Alaska Northwest Books 2002

Alaska Economic Performance Report 2002, Alaska Department of Community and Economic Development

"An Economic Engine for Alaska": Alaska Travel Industry Association (ATIA)

U.S. Bureau of Census, SER Publications

"You Decide: Where Are We Going?": Alaska 2020 Partnership Report

Interview: Dan Robinson, Statewide Labor Economist

INDEX

A

Accommodations, 248, 250
AeroMap U.S. 204-205, *204, 205*
Aerospace industry, 125-126, *126-127*,
Agricultural industry, 24, 32, *124-125*, 125
Ahtna, Inc., 44
Aialik Coast, *94-95*
Aialik Peninsula, *8-9*
Air cargo industry, 9, 121, 160-161, 164-165, 166-167
Airports, 9, 28, 63, 64, *64-65, 119*, 121, *121*, 160-161, *160, 161*, 164-165, *164-165*
Akutan, *128-129*
Alaska Aces, 103
Alaska Aerospace Development Corporation, *126, 127*, 174, *174*
Alaska Beluga Whale Committee, 51
Alaska-Canadian Highway, see also ALCAN, 64, 96
Alaska Center for the Performing Arts, 249, *249*
Alaska Center of the Performing Arts in Anchorage, *26, 75*
Alaska Department of Community and Economic Development, 122, 254-255, *254, 255*
Alaska Department of Military and Veterans Affairs-Alaska National Guard, 274, *274*
Alaska Eskimo Whaling Commission, 48
Alaska Federation of Natives, 44
Alaska Health Resources, LLC, 239, *239*
Alaska Industrial Development And Export Authority, 260-261, *260, 261*
Alaska Interstate Construction, LLC, 218-219, *218, 219*
Alaskaland, *32-33*
Alaska Native Land Claims Settlement Act, 39, 44
Alaska Native Medical Center, 228-233, *228, 229, 230, 231, 232, 233*
Alaska Native Tribal Health Consortium, 228-233
Alaska Nurses Association, The, 270, *270*
Alaska Option, 197, *197*
Alaska Peninsula, 10, 13, 28, 43, 47
Alaska Permanent Fund, 56
Alaska Purchase, 14
Alaska Railroad, 28, 55, 64, *65*, 122
Alaska Range, 24, 32, 88, 92
Alaska Regional Hospital, 238, *238*
Alaska's Flag, 67
Alaska's Marine Highway System, 96, 122, 157, *157*
Alaska State Chamber of Commerce, 264-265, *264, 265*
Alaska Statehood Act of 1958, 44
Alaska State Legislature, 256-257, *256, 257*
Alaska State Park System, 99
Alaska Travel Industry Association, 258-259, *258, 259*
Alaska USA Federal Credit Union, 194-196, *194, 195, 196*
Alaska Willow Ptarmigan, see Ptarmigan, 67
Alaska Wood Technology Center, 118
ALCAN, see Alaska-Canadian Highway
Aleut Corporation, The, 44
Aleutian Islands, *6-7*, 10, 13, 24, 28, *29*, 31, 40, 43, 47, *82-83*, 88, 125, *188-189*
Aleuts, 10, 13, 31, 40, *42*, 43, 44, 47
Alpine forget-me-not, see Forget-me-not
Alpine Pipeline, *56-57*
Alutiiq, 10, 31, 74
Alyeska Ski Resort, 99
American Seafoods Group, 178-181, *178, 179, 180, 181*
Anaktuvuk Pass, 35
Anchorage, 32, 40, 59, 63, 64, 69, 70, 73, 78, 81, 87, 88, 99, 103, 121
Anchorage Chamber of Commerce, 272, *272*
Anchorage Chrysler Dodge Center, 242-247, *242, 243, 244, 245, 246, 247*
Anchorage Fur Rendezvous, 69, *70-71, 72-73, 78-79, 102, 116, 118, 119, 215*, 251
Anchorage International Airport, see Ted Stevens International Airport
Anchorage Opera, *27, 74, 75*
Anchorage Regional Hospital, 238, *238*
Anchorage School District, 237, *237*
ANCSA, see Alaska Native Claims Settlement Act
Annette Island, 43, 140-143, *140-143*
ANWAR, 55
Apache, 10
Arctic, 35
Arctic Ocean, 10, 35, 43, 47, 84
Arctic Sea, 10
Arctic Slope Regional Corporation, 44, 208-213, *208, 209, 210, 211, 212, 213*
Art, Native, *46-47*, 47
Aspen Hotels, 250, *250*
Associated General Contractors of Alaska, 271, *271*
Athabascans, 10, 24, 31, 40, *42-43*, 43, 47, 88
Attorneys, 190-193
Auklet, Whiskered, 104
Aurora borealis, see also Northern lights, 32, *91*, 91, 96, 117
Avis Rent-A-Car Inc., 170-171, *170, 171*

Index

B
Baidarka, *42*
Banking industry, 194-196, 197, 201, 202-203
Bantam Regional Hockey, *102*
Baranof, Alexander, 13
Baranof Island, 13, *86*
Barrow, 23, 35
Barter Island, *54-55*
Basketball, 103
Bears, 27, 28, *88*, 88, 91, 92, 99, 100, *145*
Beaufort Sea, 88
Beaver, 48
Beaver Roundup, 69
Bellingham, Washington, 27
Bering Island, 10
Bering Sea, 10, 24, *31*, 43, 47, 84, 87
Bering Sea Ice Golf Classic, 77
Bering Straits Native Corporation, 44
Bering, Vitus 10
Bethel, 28
Big Delta, 117
Big Lake, 74
Biking, 103
Bison, 91
Blanket toss, *73*
Boating, 28, 95
Bristol Bay, 24, 28, 87
Bristol Bay Native Corporation, 44
British Columbia, 43, 96, 107
Brooks Range, 24, 32, 35
Building Alaska, 206-225
Business, Finance, And Professional Services, 188-205

C
Calista Corporation, 44
Camping, 99
Canada, 10, 17, 18, 24, 43, 84
Cannery, *18-19*, 59
Canoes, *18-19*, *20-21*, *44*, *48*
Cape Krusenstern National Monument, 35
Caribou, 35, 43, 48, *56*, 88, 91, *275*
Carlile Transportation Systems, 172, *172*
Cenotaph Island, *2-3*
Chief Shakes Tribal House, *30*
Chilkat Bald Eagle Preserve, 99
Chilkat State Park, 99
Chilkoot Pass, *15*
Chugach Alaska Corporation, 44
Chugach Electric Association, 162-163, *162,163*
Chugach Mountain Range, 88
Chugach National Forest, 28, 99, 100
Chugach State Park, 99
Chukchi Seas, 87, 88
Climbing, 95, *96-97*, 100
Coal, 18, 60
Cod, 28
Cold War, 31
Columbia Glacier, *22-23*, *84-85*, *148-148*

Committee on Economic Development, International Trade and Tourism, 268, *268*
Communications, 9
Construction industry, 9, 216-217, 218-219, 220-221, 222-223, 224-225
Cook Inlet, 10, 18, 23, 87, 113
Cook Inlet Region, Inc., 44
Copper, 60
Copper River, *42-43*
Cordova, 17, 28
Crab, 28, *58*, *115*
Craig, *206-207*
Crimean War, 18
Crooked Creek, 117
Crowley, 154-156, *154, 155, 156*
Cruise Ships, *62*, *63*, 96, 114, *117*

D
Dall sheep, 48, 88, 92, *97*
Dawson, 107
Deer, 40, 48, *89*, 99
Delaney, Wiles, Hayes, Gerety, Ellis & Young, Inc., 190-193, *190*, *191*, *192,193*
Delta Junction, 32
Denali, see Mt. McKinley or Denali National Park and Preserve
Denali Commission, 262-263, *262, 263*
Denali National Park and Preserve, 32, 64, 88, 92, 99

Dillingham, 13, 28, 69
Dixon Entrance, 24
Dog mushing/sledding and racing, 67, *67*, *80-81*, 81, 96, 117
Donlin Creek, 60, 117
Dowland-Bach, 222-223, *222, 223*
Doyon Ltd., 44
Dutch Harbor, 59

E
Eagle, 27, *40*, 88, 99, 100
Eaglecrest Ski Resort, 99
Education, primary and secondary, 237
Elk, 28
Endicott, *122-123*
Engineering, 9, 214-215
Eskimos, 10, 35, 40, 43, *44*, 44, 47, 48, 51, 52, 91
Evangeline Atwood Concert Hall, *26*
Export Council of Alaska, 273, *273*
Eyak Indians, 91

F
Fairbanks, 17, 23, 32, 40, 60, 64, 69, 74, 77, 99, 117, 121
Fairbanks International Airport, 121, 164-165, *164, 165*

Fairweather fault, *2-3*
Far North Region, 10, 18, 35, 40
FedEx, 166-167, *166-167*
Ferries, 27, 96, 122, 157, *157*
Fishing, Commercial, see Seafood Industry
Fishing, Sport, 28, 95, 96, 99, 100, 114
Fish rack, *48-49*
Flounder, 28
Forget-me-not, see also Alpine forget-me-not, *66*, 67
Fort Knox mine, 60, 117
Fort Richardson, 32
Four Spot Skimmer Dragonfly, 67
Fox, 99
Fur trade, 13

G

Gates of the Arctic National Park, 35
GCI, 158-159, *158*, *159*
Geophysical Institute of the University of Alaska, 91
Georgeson, Charles C., 125
Glaciers, 27, 36, *36-37*, *56-57*, 84, *84-85*, *136-137*, *138-139*, *148-149*, *226-227*, *240-241*
Gold, *15*, 17, *17*, 18, 60, 67, 117
Golden Days, 69
Golden North Salmon Derby, 69

Golden Valley Electric Association, 150-153, *150, 151, 152, 153*
Gold rush, 17, 18
Government and Community Organizations, 252-275
Great Alaska Shootout, 69
Great Gorge, *36-36*
Greens Creek Mine, 60, 117
Gulf of Alaska, 10, 17

H

H. C. Price Co., *61*, 216-217, *216*, *217*
Haida, 10, 27, 40, 43, 47
Haines, 27
Halibut, 28, 48, *147*
Harding Icefield, *240-241*
Hawaii, 84
Healy, 32
Herring, 28, 40, 48
Hiking, 95, 96, 99, 103
Historic Anchorage Hotel, 248, *248*
Hockey, *102*, 103
Homer, 59
Horizon Lines, 64, 175, *175*
Hospitals, 228-233, 234-236, 238, 239
Hubbard Glacier, *176-177*
Huna Totem, 173, *173*

I/J

Icy Bay, 24
Iditarod Trail Sled Dog Race, 69, 70, 77, *80-81*, 81,
Iliamna, 117
Illinois Creek mine, 118
Inside Passage, 27, 40, 87, 96
Interior Region, 10, 17, 24, 28, 31-32, 43, 47, 60, 64, 74, 117, 118
International Whaling Commission, 48
Inupiat, 10, 35, 40, 47
Isanotski Peaks, *6-7*
Jade, 67
Johns Hopkins Glacier, *226-227*
Johnson, President Andrew, 14
Juneau, City and Borough of, 14, 17, *34*, 60, 67, 69, 73, 96, 99, 117, *117*, 136-139, *136-139*
Juneau Gold Rush Days, 73
Juneau International Airport, 121

K

Kachemak Bay State Park, 99
Kasegaluk Lagoon, 87
Katmai National Park and Preserve, 24
Kayaking, 28, 95, 103
Kenai Peninsula, 27, 28, 125, 126
Kensington/Jualin Mine, 117

Ketchikan Gateway Borough, *30*, 59, 60, 118
Ketchikan International Airport, 121
Killer whale, see Orca
King Salmon, City of, 28
Kipnuk, *38-39*
Klondike Gold Rush, 18
Klondike Kate, 107, *107*
Klondike River, 18
Knik Arm Bridge, 55
Kobuk Valley National Park, 35
Kodiak Crab Festival, 69, 74
Kodiak Island, 10, 13, 24, 28, 31, 59, 74, 84, 87, 125, 126
Kodiak Launch Complex, 126, 174, *174*
Koniag, Inc., 44
Kotzebue, 23, *24-25*, *87*, 117
Kougarok, 117
Kugkaktlik River, *38-39*

L

Law firms, 190-193
Lead, 60
Leathard, Peter, *9*
Liquid natural gas, 56
Little Norway Festival, 69, *76*, 78
Little Port Walter, *86*
Lituya Bay see also *Port des Francais*, *2-3*, 43
Lodges, *101*
Lund, Ethel, *39*

M
Mackerel, 28
Magnet, The, *12-13*
Marine Stewardship Council's Sustainability Seal, 59
Marketplace, 240-251
Marmot, *70*
Masks, *46-47*, 47
Matanuska-Susitna Valley, 27, 125
Mt. McKinley, 32, *36-37*, 88, 92, *92-93*, 100, *252-253*
Medical, 228-233, 234-236, 238, 239
Mendenhall Glacier, *136-137*
Metlakatla, 23, 43, 140-143, *140-143*
Miners, *15*, 17
Mining, 6, *32-33*, 56, 60, *60*, 110, 117, 182-183
Moe, Tommy, 99
Moose, 48, *66-67*, 67, 88, 91, 92, 99, 100
Moose's Tooth, *68-69*
Mountain climbing, 100
Mountain goat, 31, 48, 91, 99
Mt. Crillion, *2-3*
Mt. Huntington, *148-149*
Mt. Marathon Race, *103*
Mt. St. Elias, 88
McKinley, William President, 88
Murkowski, Frank H., *6*
Musk Ox, *52*, 52

N
N.C. Machinery Company, 220-221, *220*, *221*
Naknek, 28
NANA Region Corporation, Inc., 44
National Marine Fisheries Service, 51
National Oceanic and Atmospheric Administration, 31, 86, 87
Native art, *46-47*, 51, 52, *52-53*
Native corporations, 28, 118, 173, 208-213
Native culture, 51, 52
Native Youth Olympics, 78
Natural Resources, 176-187
Nature Conservancy, The, *110*, 266-267, *266*, *267*
Navajo, 10
Nelson, Valorie, *87*
Nenana, 81
Nenana River, 74
Nenana River Ice Classic, 74
Networks, 150-175
New Archangel Bay, 13
New York, New York, 107
Nixon, President Richard, 44
Noatak National Preserve, 35
Nome, *14*, *15*, 17, 18, 23, 70, 74, 77, 81

Nome Gold Rush, *14*, *15*, 17
Northern lights, see also aurora borealis, 32, *91*, 91, 96, 117
North Pole, Alaska, 64
North Slope, 24, 32, 35, 56, 110
Northwest Coast Cultures, 10, 40
Norton Sound, 87

O/P
Oil and gas development, 6, 18, 28, *54-55*, 56, *108-109*, *110-111*, 110-113, *112-113*, *120-121*, *122-123*, 184-185, *184*, *185*, 186-187, *186*, *187*, 214-215, 216-217, 218-219, 220-221, 222-223
Okmok Caldera, *188-189*
Oomiak, *44*, *50-51*
Orca, *84*, 88
Owl, *98*
PGS Onshore, 184-185, *184*, *185*
Paddleboat, *33*
Parka, *50*
Pavlof Volcano, *29*
Paxon, 117
Peak Oilfield Service Company, 186-187, *186*, *187*
Pearce, Drue, *85*
Pebble Copper project, 117
Petersburg, *1*, 59, 69, 76, *78*, 99, *114*
Peter the Great, 10
Petroglyph Beach State Park, *144*

Pignalberi, Mary, *69*
Placer, 60
Pogo gold project, 117
Point Barrow, 91
Point Bridget State Park, 99
Poker Flat Rocket Range, 125-126
Polar Bear Jumpoff Festival, 74
Pollock, 28
Porpoise, 91
Port des Francais, see also Lituya Bay, *2-3*, 43
Port of Anchorage, 28, 55, 64, *64*, *120*, 122, *122*, 168-169, *168*, *169*
Port of Whittier, 122
Potlatch, *45*
Power companies, 162-163
Premera Blue Cross Blue Shield of Alaska, 198-200, *198*, *199*, *200*
Pribilof Islands, 28, 31, 43
Prince of Wales Island, 10, 43
Prince William Sound, 17, 27, 28, 100, 122, *148-149*
Providence Health System, 234-236, *234*, *235*, *236*
Prudhoe Bay, 35. 110
Ptarmigan, see also Alaska Willow Ptarmigan, 67
Ptarmigan Lake, *98-99*

Q/R

Qiviut, 52, *52-53*
Quality Of Life, 226-239
Quiviut, see Qiviut
Rafting, 95
Ragged Jack, *6-7*
Railroad, *16*, 17, *17*
Red Dog Mine, 60, 117
Red Dog Saloon, *34*
Reeder, Fred, 107
Reflection Lake, 92
Resource Development Council, 269, *269*
Resurrection Bay, 74
Rockets, *126*, *127*, *174*
Rockfish, 28
Rockwell, Kathleen Eloisa, 107, *107*
Russian American Company, 13
Russian Period, 10-13
Russian-Tlingit War, 13
Ruth Glacier, *36-37*
Ryan, Katherine, 107

S

Sablefish, 28
Salmon, 27, 28, 40, 48, 67, 91, 99, 100
San Francisco, 21
Scenic Byways, 28
Seafood Industry, 6, 18, *18-19*, 56, *58-59*, 59, 110, 113-114, *114*, *115*, 178-181
Sealaska Corporation, 44, 118
Seal/sea lion, *24*, 48, 91, 100, *110*
Sea otters, 91, 99
Seattle Mortgage, 201, *201*
Seattle, Washington, 18
Selawik National Wildlife Refuge, 35
Seward, City of, 17, 64, 74, 87, 121, 146-147, *146-147*
Seward Highway, 28
Seward Peninsula, 117
Seward's Day, 14
Seward's Folly, 14, 18
Seward's Icebox, 14, 18
Seward, William H., *12*, 14, 18, 84
Sheffield, Governor William J., 55
Shelikof, Gregory, 13
Shishaldin Volcano, *82-83*
Shuyak Island State Park, 99
Siberia, 40
Silver, 60
Sitka, City and Borough of, 13, 14, 23, 45, 59, 60, *79*, *109*, 130-135, *130-135*
Sitka Spruce, 67
Skagway, City of, *16*, 18, 104
Skiing, 96, 99, 100, 103, 117
Smith, Jefferson "Soapy", 104, *104-105*
Snowboarding, *119*
Snow machines, *72*, 95, 99, *102*, 117

Snowplow train, *backcover*, *10-11*, *17*
Snowshoeing, 99
Sourdough, 21, *21*
Sourdough bread, 21
Southcentral Region, 24, 27-28, 118, 125
Southeast Region, 10, 24-25, *35*, 43, 47, 60, 84, 87, 88, 91, 99, 117, 118, 122
Southwest Region, 24, 28, 31, 40, 47
Spokane, Washington, 107
Staser, Jeff, *23*
Statehood, 14, *79*
State symbols, 67
St. Elias Mountain Range, 88
St. George Island, 43
St. Paul Island, 28, 43
Subsistence lifestyle, 48, 51
Sugiaq, 10, 40
Superior Plumbing & Heating, Inc., 224-225, *224*, *225*

T

Taiga, 32
Talkeetna Moose Dropping Festival, 69
Tanana Valley, 125
Teck Cominco Alaska Inc., 182-183, *182*, *183*
Ted Stevens Anchorage International Airport, 28, 63, 64, *64-65*, *119*, 121, 160-161, *160*, *161*
Telecommunications, 109, 158-159
Tesoro Iron Dog Snowmachine Race, 74
Thirteenth Regional Corporation, 44
Three Saints Bay, 13
Timber Industry, 6, 56, 60, 110, 118
Tlingit, 10, 13, 27, 39, 40, *43*, *45*, 47, 91
Tongass Land Use Management Plan, 60
Tongass National Forest, 24, 27, 28, 60, 99, 118
TOTE, 64
Totems, *40-41*, 43, *46*, *47*, 47, *140*, *141*, *173*
Tourism Industry, *62*, 63, *63*, 96-103, *101*, 114-117, 258-259
Trains, *10-11*, *16*, *17*
Tram, *62*
Trans-Alaska Pipeline, 35, *61*
Transportation industry, 9, 63, 154-156, *154*, *155*, *156*, *172*
Treaty of Cession 1867, *13*
True North Mine, 117
Tsimshian, 27, 40, 43, 47
Tundra, 24, 32, 35, 52
Turnagain Arm, 87

U/V

U.S. Census Bureau, 18
U.S. Fish and Wildlife Service, 275, *275*
Umnak Island, *188-189*
Unalaska, 13, 24, 28
Unangon, 10
Unimak Island, *6-7, 82-83*
Union Bay Mine, 117
University of Alaska Anchorage, 103
University of Alaska Anchorage Seawolves, 103
University of Alaska Fairbanks, 32, 125, 126
Valdez, City of, 17, 23, 28, 35, 100
Vaughan, Norman, *95*
VECO Corporation, 214-215, *214, 215*
Volcano, *6-7*, 28, *29*, *82-83, 188-189*

W

Walrus, 48, 74, *100*
Washington, state of, 96
Wasilla, 81
Wells Fargo Bank Alaska, 202-203, *202, 203*
West Coast Hockey League, 103
Western Collegiate Hockey Association, 103
Whales, 48, *84*, 87-88, 91, 99
Whaling, 48, 51
Whitefish, 48
Whitehorse, 18
Whittier, City of, 28, *28*, 64, 121
Windham Bay, 17
Windsurfing, 97
Wolf, 88, 92
Wood-Tikchik State Park, 99
Wooley Mammoth, 67
World Extreme Skiing Championship, 100
World Ice Art Championship, 77
World War II, 31, 96
Wrangell, City of, *30*, 144-145, *144-145*
Wrangell Mountain Range, 88
Wrangell-Saint Elias National Park and Preserve, 88

Y/Z

Yakutat, 13, 24
Yukon-Kushokwim Delta, *38-39*
Yukon River, 18, *20-21*, 31, 84
Yukon Territory, 17, 18
Yun-day-stuck-e-yah Village *18-19*
Yup'ik, 10, 24, 31, 40, 47
Zinc, 60

Ms. Megan Hatch, a student at Service High School, created the dust-jacket illustration. She won the 2002 Wyndham Publications-Anchorage School District Cover Art Contest for the *Alaska: North to the Future, Volume II* project.